ENERGY SAVING BY INCREASING BOILER EFFICIENCY

ENERGY SAVING BY INCREASING BOILER EFFICIENCY

Edited by L.H. Yaverbaum

NOYES DATA CORPORATION
Park Ridge, New Jersey, U.S.A.
1979

Library of Congress Catalog Card Number: 79-83772
ISBN: 0-8155-0745-3
Printed in the United States

Published in the United States of America by
Noyes Data Corporation
Noyes Building, Park Ridge, New Jersey 07656

Foreword

Boiler plant managers and operators of single boilers will find that this definitive text furnishes the precise assistance needed to implement efficiency improvement techniques. With the careful boiler maintenance described here fuel savings up to 30% can be realized.

To keep efficiency high, each device in the system must be adjusted and tuned for optimum performance. This book presents the necessary information regarding adjustments, modifications, equipment and maintenance that should be considered, from the simplest adjustment of the fuel/air ratio (which incidentally can effect remarkable fuel savings of several percentage points) to the installation of more complex auxiliary equipment. There is included instructive material on installing combustion controls, minimizing heat losses, adjusting combustion conditions, and removing impurities.

Not to be disregarded are the data on alternate fuels. The testing and monitoring techniques so essential to better functioning of the boilers are also discussed in detail, and in the last chapter the economic effects of suggested changes are evaluated.

Because the information in this book is taken from multiple sources, it is possible that certain portions of this book may disagree or conflict with other parts of the book. This is especially true of monetary values and opinions of future potential. We chose to include these different points of view, however, in order to make the book more valuable to the reader.

Cost figures provided are those given in the report cited, the date of which is always given. When the dates of the cost figures themselves are given, we have included them.

Advanced composition and production methods developed by Noyes Data are employed to bring these durably bound books to you in a minimum of time. Special techniques are used to close the gap between "manuscript" and "completed book." Industrial technology is progressing so rapidly that time-honored, conventional typesetting, binding and shipping methods are no longer suitable. We have bypassed the delays in the conventional book publishing cycle and provide the user with an effective and convenient means of reviewing up-to-date information in depth.

The expanded Table of Contents provides easy access to the information contained in this book which is based on several studies produced under contract for the Federal Energy Administration, the U.S. Department of Energy, and the U.S. Environmental Protection Agency. These primary sources are listed at the end of the volume under the heading Sources Utilized. The titles of additional publications pertaining to topics covered in this book are listed at the end of most chapters and some sections.

Some of the illustrations in this book may be less clear than could be desired; however, they are reproduced from the best material available to us.

Contents and Subject Index

Introduction

The threat of fuel shortages is plaguing the nation and fuel oil prices have surged these last few years. Therefore, it is imperative that ways of conserving fuel be found. Since industrial boilers currently account for about 17% of all the energy consumed in the United States (not including the steam or hot water generated in schools, hotels, commercial buildings, electric power plants, etc.), or more than eleven quadrillion Btus per year, improving boiler efficiency is of prime concern. An increase in industrial boiler efficiency would significantly impact energy conservation nationwide, as well as reduce the fuel costs of industry.

A fuel-burning system can be viewed as an interconnection of devices which prepares fuel and air for combustion, directs their introduction into the furnace, provides for stable ignition and combustion, and suitably removes the products of combustion from the furnace. When fuels were abundant and lower in cost, maintenance was frequently postponed; yet these systems must be maintained and adjusted to keep fuel efficiency high. When equipment is properly adjusted and maintained, fuel savings of from 10 to 30% can be obtained. Field test programs have revealed that efficiency improvements of up to several percent are frequently possible simply by adjusting the fuel/air ratio to minimize stack gas heat losses.

This book attempts to assist in the evaluation of efficiency improvement techniques for industrial boilers. It supplies information for plant managers, owners, and engineers, as well as boiler operators. It is imperative that boiler operators be fully cognizant of all aspects of the operation. The operator is vital to the success of the operation; all pertinent information regarding adjustments, modifications, equipment, and maintenance should be transmitted to him accordingly.

Proponents of fuel additives feel that additives can boost burning efficiency, eliminate corrosion, solve slagging and pollution problems, or prevent sludge buildup; however, they are discussed in another book in this series entitled *Boiler Fuel Additives for Pollution Reduction and Energy Saving* (1978).

Among the measures for limiting fuel consumption described in this book are installation of combustion controls, heat loss minimization, adjustment and maintenance of burner and furnace conditions, removal of impurities inherent in the system, addition of auxiliary equipment, efficiency calculations and alternative fuel analyses, as well as the economic effects caused by these changes.

It is hoped that with a better understanding of all aspects of boiler operation, maximum use of each Btu of energy will be obtained. The ultimate objective is to stimulate the use of the most energy-efficient system economically practicable, thereby accomplishing a significant conservation of fuel. These efforts are not only important to the individual user, but are also vital to the national goals of energy conservation.

Overview of Boiler Efficiency and Controlling Factors

The material in this chapter has been based upon a report by KVB, Inc. (PB-262 577).

This chapter gives a brief description of important factors affecting or limiting boiler efficiency and lists the more common operational modification and efficiency improvement equipment alternatives that are available. Boiler efficiency is a measure of the efficiency with which the heat input to the boiler (principally the higher heating value of the fuel) is converted to useful heat output (in the form of process steam).

MAJOR FACTORS AFFECTING BOILER EFFICIENCY

Improvements in steam generator efficiency result primarily from reductions in waste heat energy losses in the stack gases and expelled water. Procedures that reduce the mass flow and energy content of these flow streams directly benefit unit performance. Other losses occur from surface heat transfer to the atmosphere and incomplete combustion of the fuel.

Overall boiler efficiency is often expressed as 100% minus the total losses (in percent). Maximum efficiency results when the combustion of the fuel is complete and the heat losses described above are minimized. The direct approach to improving efficiency is to identify the losses, their relative magnitude, and then concentrate first on the dominant losses that are controlling the degraded efficiency.

Heat Loss Minimization

Waste heat energy losses in the stack gases consist of the dry flue gas loss (heat carried away by the dry flue gases) and the moisture loss (latent and sensible heat in water vapor). Water vapor results from the combustion of hydrogen in the fuel, the humidity of the combustion air, and the water content of the fuel.

Most industrial boilers have very large dry flue gas losses because they operate with high excess air levels and high stack gas (exit) temperatures. Traditionally, these boilers have not had the sophisticated combustion controls common on larger utility boilers and as a result the furnace is operated at very high excess air levels to insure complete combustion and safe operation. These high excess air levels result in the unnecessary heating of tons of oxygen and nitrogen that are subsequently dumped into the atmosphere.

The latent heat of water vapor usually comprises a large fraction of the total efficiency losses and could be recovered if the water vapor were allowed to condense out before the flue gases leave the boiler. Since this would be impractical in conventional boiler systems, the latent heat is not considered recoverable.

In examining efficiency losses, it is apparent that any reduction in the exit flue gas temperature will help minimize the flue gas losses. Minimum flue gas exit temperatures are limited by corrosion and sulfuric acid condensation in the cold end regions of the unit. Because of the importance of reducing flue gas exit temperature, many operational and equipment modifications are concerned with means for accomplishing this reduction.

Combustion Efficiency

Another source of efficiency loss in an industrial boiler that is not operating properly or is poorly maintained is the loss due to incomplete combustion. This includes the loss due to combustible material in the flue gases (carbon monoxide, hydrogen, carbon carryover, hydrocarbons, and smoke), and the loss due to unburned solid fuel which becomes trapped in the refuse. As previously mentioned, the high excess air levels frequently used in industrial boilers tend to minimize this loss unless the boiler is improperly maintained, an older poorly designed coal unit, or burning an uncommon fuel of inconsistent quality.

The optimum excess air level for the best boiler efficiency occurs when the sum of the loss due to incomplete combustion and the loss due to heat in the flue gases is a minimum. For the ideal case of rapid thorough mixing, the optimum air-fuel ratio is the stoichiometric air-fuel ratio. However, excess air is required in all practical cases to increase the completeness of combustion, allow for normal variations in the precision of combustion controls, and insure satisfactory stack conditions with some fuels (i.e., nonvisible plume to comply with air pollution regulations). The optimum excess air level will vary with fuel, burner, and furnace design and can be determined by tests conducted with different air-fuel ratios.

Other Factors

Heat lost from the boiler jacket through its insulation is generally termed "radiation loss" and includes heat radiated to the boiler room and the heat picked up by the ambient air in contact with the boiler surfaces. The quantity of heat lost in this manner (Btu/hr) is fairly constant at different boiler firing rates and as a result becomes an increasingly higher percentage of the total heat losses at the lower firing rates. To a large extent these losses are unavoidable but deteriorated insulation and furnace wall refractory will increase these losses at all loads.

Wastewater losses occur from excessive drum water blowdown beyond that required to maintain satisfactory dissolved solids concentration and from discarded recoverable condensate. These represent not only lost energy but also a wastage of water and the chemicals used in its treatment. The increasing cost of energy makes the recovery of these losses more economical.

The degree to which operational and equipment modifications can benefit industrial boiler performance is determined by the type and condition of the steam generating unit including the firing system and combustion controls, the fuel employed, the existing heat recovery equipment and the general efficiency related operating practices. A complete engineering evaluation from the standpoint of economics, system capability and practicality is required prior to the selection of modified operating practices or auxiliary equipment installation.

BOILER MAINTENANCE PRACTICE AND OPERATIONAL MODIFICATIONS

The previous discussion has indicated that efficiency improvement can be accomplished both through improved operating practices and by purchasing efficiency improvement equipment. In evaluating these options, it is important to realize that efficiency tests at industrial boiler installations (1)(2) have demonstrated that in many cases substantial improvements in day-to-day boiler operating efficiencies can be readily achieved without requiring the purchase and installation of additional heat recovery equipment, new boiler controls or other retrofit devices.

In particular, modified operating practices and proper maintenance of efficiency-related hardware can minimize efficiency losses and lead to the most efficient utilization of existing steam generating equipment. The return on investment with these techniques is usually very high and may offer an attractive alternative to installing more costly retrofit devices where the installation of these devices cannot be afforded or economically justified. Even when retrofit devices are used, the implementation of these operating and maintenance practices would assure that the maximum efficiency potential of the modified boiler is achieved and continually maintained.

There are methods or techniques which concern the more promising changes in boiler operating pactice that reduce fuel consumption. These include not only means for directly improving boiler efficiency but other techniques such as load management where fuel is saved by distributing load between boilers so that each boiler is operating near its peak efficiency. Generally, any operating change that has a beneficial effect on stack gas flow or temperature, combustibles in the fly ash and flue gas and wastewater losses is efficiency or fuel conservation related.

An important factor in preserving these benefits is careful attention to boiler maintenance items. In addition to maintaining efficient boiler operation, these factors generally have an important bearing on plant reliability, safety and load-carrying ability.

Operating Practices

Boiler Maintenance: These practices are important in achieving or maintaining efficient operation. Poor maintenance resulting in degraded equipment operation in a manner that affects stack gas temperature, excess air level, combustibles, or other heat losses, can affect boiler efficiency and fuel consumption.

Boiler Tune-up: This is one of the most cost-effective means of achieving efficient operation and saving fuel. Adjustment and maintenance of fuel burning equipment and combustion controls permits operation with the lowest practical excess air, thus reducing stack losses. Annual tune-ups can be performed by boiler operating personnel, service organizations, some local utilities, burner/boiler manufacturers, or engineering consulting firms. A routine tune-up on a 25,000 lb/hr boiler can typically be conducted by a burner manufacturer in a day or two at $300/day plus expenses.

Boiler Tube Cleanliness: Cleanliness is important for efficient heat transfer from the hot combustion gases to the water side passages. Tube deposits and fouling on external tube surfaces of a watertube boiler will be reflected in high flue gas temperatures. An estimated 1% efficiency loss occurs with every 40°F increase in stack temperature. Some boiler operators perform their own tube washes while others rely on commercial cleaning and repair service firms that can charge as low as $200 to clean a 10,000 lb/hr unit. The actual cost depends on the size of the unit, fuel type, furnace design and extent of deposits.

Reducing the Boiler Steam Pressure: Where permissible this is an effective means of reducing fuel consumption by as much as 1 to 2%. Lower steam pressures give a lower saturated steam temperature, and without stack heat recovery a similar reduction in final flue gas temperature will follow. These benefits can be achieved at little or no expense providing the process or equipment can accommodate lower steam pressures. This is especially applicable to boilers that have been converted to hot water heaters.

Blowdown: This is a customary procedure to remove boiler water with high impurity concentrations. It also represents a loss of sensible heat energy in the wastewater. Improved blowdown practices or the use of automatic equipment can result in a fuel savings dependent upon current practice.

Load Management: To minimize fuel consumption in a plant with several boilers load management is used. Since boiler efficiency varies with load, boiler design, equipment age, fuel, and other factors, there is an advantage to distributing load to the most efficient boilers and in operating boilers at loads where efficiency is highest. There also is merit to shutting some boilers down in swing seasons while operating process boilers at peak efficiency as opposed to operating all boilers at reduced load and performance.

Fuel Change: Fuel conversion or fuel change is generally not thought of as a common means for fuel conservation since the decision usually is controlled by availability. However, different fuels do characteristically exhibit modest differences in boiler efficiency which should be taken into account in addition to raw fuel cost on a Btu per dollar basis.

BOILER AUXILIARY EQUIPMENT FOR EFFICIENCY IMPROVEMENT

The descriptions above have briefly summarized methods of efficiency improvement and fuel conservation that can be accomplished with little or no capital expenditures. Although these methods are very cost-effective, major improvements in efficiency can be accomplished more readily with equipment modifications that might require several years for return on investment.

This section reviews the more commonly used efficiency improvement equipment to describe the range of options from which one can choose and the relative costs and efficiency improvement that can be expected.

There are various off-the-shelf items that can be employed to improve boiler performance. These items can generally be divided into heat traps that lower the temperature of the stack gases, low excess air equipment that reduces stack gas flow, and other items that limit waste heat and radiation losses.

The efficiency improvement potentials are not cumulative, that is, a combination of a heat trap and a low excess air modification will not result in an efficiency improvement that is the sum of each individual item acting alone.

Boiler selection should also be considered if it is shown to be more cost-effective to purchase a new boiler unit than utilize auxiliary equipment on an existing unit.

Equipment Synopsis

Air Preheaters: These are heat traps installed in the exhaust duct that transfer heat energy from the hot flue gas to the incoming combustion air supply. Unit efficiency will increase approximately 2.5% for every 100°F decrease in flue gas temperature. Additional performance benefits will result from improved combustion conditions in the furnace which may enable operation with lower excess air. Application is limited by corrosion considerations (i.e., lowest acceptable stack temperature), existing stack temperatures, maximum acceptable combustion air temperature and associated increases in NO_x emission levels. Initial capital equipment costs range from $1,000 to $2,000 per MBtu/hr, depending on boiler capacity.

Economizers: These are heat traps that preheat boiler feedwater using heat energy from the flue gas. Again, unit efficiency will improve 2.5% for every 100°F decrease in flue gas temperature. Corrosion considerations are the primary limitation with some consideration required for boiler operating pressure on low pressure boilers.

Economizers are generally preferred over air preheaters on small units. A combination of air preheaters and economizers is employed on large units with operating pressure above 400 psig. Economizer initial costs are comparable to that of air preheaters but other factors of installation and operating cost are a consideration.

Turbulators: Baffles installed in the secondary passes of a firetube boiler to increase turbulence and thereby convection heat transfer to the surrounding boiler water are called turbulators. As with the heat traps mentioned above, the improvement in boiler performance will be indicated by the decrease in stack gas temperature and the corresponding increase in steam generation. Turbulators cannot be used on coal fired units. Turbulator installations range from $500 to $2000 per unit, depending upon unit size, and are economically attractive to many firetube boiler operators.

Improved Combustion Controls and Instrumentation: These are designed to reduce the operating excess air levels required for safe operation. The degree to which boiler performance is improved will depend on the previous stack gas excess air and temperature conditions and the reduced excess air and temperature possible with new controls and instrumentation. A changeover from the simplest jackshaft system to a cross-limited excess oxygen system can improve boiler performance by 2% and 5% for systems with and without heat traps respectively. Approximate installed costs range from $30,000 to $150,000, depending on the degree of sophistication and capacity of the unit.

Low Excess Air Burners: These burners achieve complete combustion on oil and gas fuel with as little as 1-3% excess air. Associated decreases in NO_x and particulate emissions have also been reported. Sustained operation at such low excess air levels however is questionable and will require a very sophisticated combustion control system.

Initial capital costs range from \$700 to \$200 per MBtu/hr depending on the size of the unit.

Packaged burner assemblies that come equipped with their own control system can reportedly operate with 2-3% excess O_2 levels and are available for retrofit replacements of the existing burner and control system. These are generally priced between \$10,000 and \$35,000 for a single burner unit.

Wastewater Heat Recovery: This involves the recovery of waste heat energy contained in drum water blowdown and expelled condensate. This heat energy can be effectively used to preheat boiler feedwater instead of just discarded. Approximately a 10°F increase in feedwater temperature will result in a 1% improvement in efficiency. Nominal costs for automatic blowdown systems and condensate recovery units are \$1,000 and \$8,000, respectively.

Insulation: Insulation will reduce radiation and convective heat losses from the boiler jacket and ductwork to the boiler surroundings. An estimate of these energy losses can be made based upon surface temperatures, air flow and surrounding conditions, and should be made prior to the purchase of new insulation where improvements appear marginal. The cost of improved insulation is highly variable and is a function of material used, desired skin temperature, and condition of existing insulation.

Sootblowers: These are used to remove soot, slag and fly ash deposits from the tube surfaces that would otherwise retard heat transfer in the convective regions leading to higher stack temperatures. Manufacturers claim that up to a 1% efficiency improvement can be achieved by maintaining clean tube surfaces. A simple sootblower system will cost \$15,000 to \$50,000.

REFERENCES

(1) G.A. Cato, et al, *Field Testing: Application of Combustion Modifications to Control Pollutant Emissions from Industrial Boilers—Phase I,* KVB, Inc. EPA 650/2-74-078a, Oct. 1974.

(2) G.A. Cato, et al, *Field Testing: Application of Combustion Modifications to Control Pollutant Emissions from Industrial Boilers—Phase II,* KVB, Inc. EPA 600/2-76-086a, Apr. 1976.

Combustion in Boiler Systems

The material in this chapter has been based upon a report by KVB, Inc. (PB-264 543) and another report by Engineering Extension Service, Auburn University (PB-265 713).

FUNDAMENTALS OF COMBUSTION IN BOILER SYSTEMS

The heat energy in all fossil fuels (coals, fuel oils, and natural gases) is released during the combustion process as the carbon and hydrogen in the fuel react with oxygen to produce a high temperature flame. The burning of carbon and hydrogen with oxygen would ideally produce an exhaust gas consisting of carbon dioxide and water vapor which are considered harmless combustion by-products.

Unfortunately, real combustion systems differ from this ideal situation due to the more complex make-up of the fuel itself plus the nonideal characteristics of the actual burning process. As a result, the exhaust products from industrial boilers contain additional gaseous and solid materials, some of which are identified as serious environmental pollutants. Some are products of incomplete combustion and are important from an efficiency standpoint, since they represent a waste of available heat that is lost from the boiler stack.

Air is the convenient and nonvarying source of oxygen for nearly all industrial combustion processes. The composition of air is roughly 21% oxygen and 78% nitrogen by volume with traces of other gases including argon and carbon dioxide. Under perfect combustion conditions, there is a so-called "theoretical amount" of air which will completely burn a given amount of fuel with no excess air remaining. In practice, however, more air than the theoretical amount must be supplied to the burner to ensure complete burning of the fuel. The actual quantity of excess air required at a particular boiler depends on many factors such as fuel type and composition, furnace design, boiler firing rate, and the design and adjustment of the burners. If enough air is not provided, unburned fuel, smoke and other products of incomplete combustion such as carbon monoxide and soot are emitted from the stack, resulting in pollution, wasted fuel energy, and a potential for explosion.

In the case of coal fired boilers, the quantity of unburned fuel in the refuse also increases. If too much excess air is supplied to the burner, the boiler efficiency will be reduced because some fuel is used simply to heat the unneeded air which is exhausted out the stack. The amount of excess air can also influence the formation and emission of nitrogen oxides (NO_x).

Furnace Excess Air

Excess air is a primary boiler operating variable and an important element in this discussion. Accurate fuel and air flow measurements would provide a direct determination of the boiler air/fuel ratio and excess air, but these measurements are frequently not available at industrial boiler installations. An alternate means for determining boiler excess air utilizes measurements of specific gases in the stack and the known relationships between their concentration and percent excess air for a particular fuel.

Excess air in the boiler is evident from the presence of oxygen in the flue gas. This oxygen content is customarily referred to as excess oxygen, excess O_2 or stack O_2. However, the terms "excess air" and "excess O_2" should not be confused.

As the excess air is increased at the boiler, one would expect that the oxygen concentration in the stack gases would also increase, and, indeed, this is the case. Figure 2.1 shows this relationship between excess O_2 and excess air for typical natural gas, oil and high grade coal fuels.

FIGURE 2.1: BOILER EXCESS AIR vs STACK GAS CONCENTRATIONS OF EXCESS O_2 AND CO_2

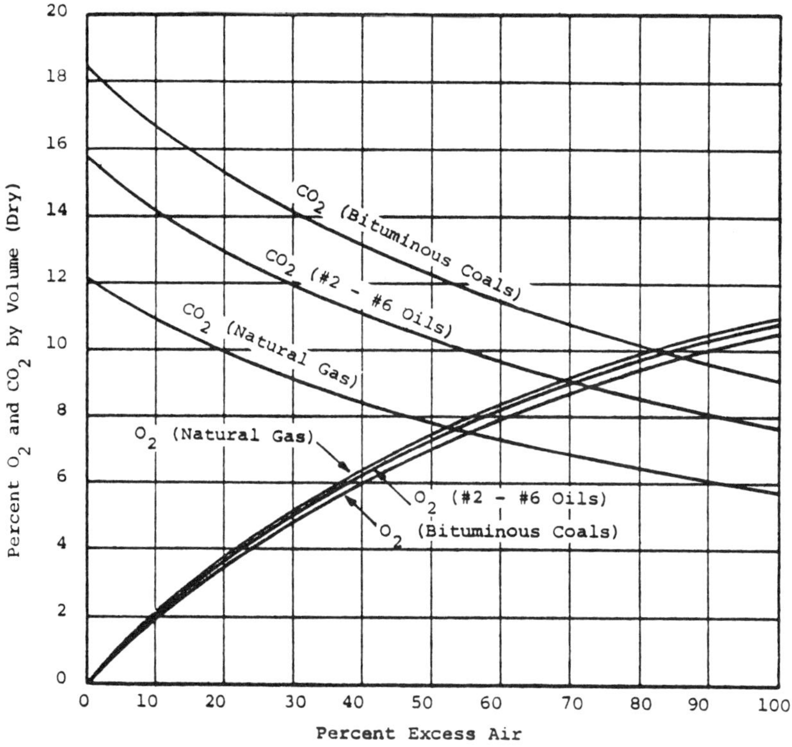

Source: PB-264 543

Some boiler operators judge boiler firing conditions based on stack CO_2 measurements (as opposed to using excess O_2) and the CO_2 versus excess air dependence is also shown in Figure 2.1. While both approaches are equally valid for determining boiler excess air, the use of excess O_2 is preferred for the following reasons:

1) The relation between O_2 and excess air is not highly affected by fuel composition. This is evident in Figure 2.1 where the O_2 vs excess air curves for natural gas, oil and coal fuels are nearly coincident. On the other hand, CO_2 vs excess air curves exhibit a wide variation with fuel type.

2) Measurement of CO_2 requires a much greater precision than excess O_2 measurements to obtain the same accuracy in excess air. This is especially true when firing with low excess air which is generally the preferred condition for high boiler efficiencies. For example, suppose it is desired to establish the excess air to within $\pm$2% excess air. If the boiler is operating in the region of 20% excess air on oil fuel, the CO_2 would have to be measured to within $\pm$1.5% accuracy (13.0$\pm$0.2% CO_2), while an O_2 measurement accuracy of only $\pm$8% (3.7$\pm$0.3% O_2) would be required.

3) CO_2 is a product of combustion whereas excess O_2 is more immediately associated with excess air conditions in the boiler. The excess O_2 measured in the stack gases is the actual surplus of oxygen in the furnace available for the combustion process. As lower and lower excess air is present in the furnace, the excess O_2 approaches zero (regardless of fuel type) while the CO_2 increases, approaching some constant value which is dependent on the fuel composition.

4) Instrumentation for excess O_2 measurements is generally less expensive and more reliable as a class than CO_2 instrumentation. Many continuous CO_2 instruments are of a NDIR type which is more expensive and not very portable.

In the following discussion, excess O_2 will be used exclusively as an indicator for excess air. It is strongly recommended that the use of excess O_2 be adopted by all boiler operators. However, for those who use CO_2, it is possible to convert O_2 to CO_2 using curves such as those in Figure 2.1 for the particular fuel being fired.

The interpretation of boiler excess air as it relates to excess air at the burner should also be noted. For boilers with a single burner, the percent excess air at the burner is said to be the same as the boiler excess air (percent excess air at the stack).

For multiburner boilers, the situation can be complicated if the fuel and air are not equally distributed to all the burners. With equal fuel and air distribution, the fuel/air ratio and percent excess air at all burners is the same and equal to the percent boiler excess air. However, when the fuel and/or air varies from burner to burner (due to dissimilar air register settings, eroded fuel injection orifices, coal pipe obstructions, etc.), the fuel/air ratio and burner excess air will also vary. The use of "overfire" air or other air injection ports away from the burner region will of course result in less excess air at the burner than indicated by the stack O_2. This situation generally occurs only with stoker fired boilers where the concept of burner excess air is not usually applicable due to the nonuniform or time-dependent nature of solid fuel bed burning. In this case, stack excess O_2 is still an important measurement.

Casing or setting air leaks on induced or balanced draft units can also distort the O_2 readings. In these cases comparison of overfire O_2 and stack O_2 readings can verify the presence of air infiltration to the furnace.

Nitrogen Oxides

Combustion temperatures on the order of 3500°F are reached in the flame zone of industrial boilers. When oxygen and nitrogen are both present at such high temperatures, a fraction of these will combine to form NOx. Nitrogen oxides formed in this manner are difficult to prevent altogether, since all three ingredients (oxygen, nitrogen and high temperature) are integral parts of the combustion process. Nitrogen is the major constituent in the combustion air and its elimination would be very impractical. Most NOx reduction techniques are generally viewed to be effective as a result of lower peak flame temperature, reduced availability of oxygen in the flame or a combination of both. Oxygen and temperature are important flame parameters affecting stability, radiant heat release, combustible burnout, flame appearance and other boiler operation related factors. Therefore, the successful application of NOx reduction techniques involves more than just reducing NOx emission levels.

Equal consideration must be given to these other factors to assure safe, reliable and efficient operation. This aspect of implementing low NOx firing will involve a certain amount of subjective interpretation of flame conditions in the furnace coupled with an evaluation of boiler efficiency, steam conditions and other boiler operational factors.

Nitrogen oxides produced by the combining of atmospheric nitrogen and oxygen under the influence of high flame temperatures are generally referred to as "thermal NO_x." Conventional solid and liquid fuels (even some natural gas fuels) contain quantities of nitrogen bearing compounds. A fraction of this nitrogen also combines with oxygen in the flame to produce what is called "fuel nitrogen NO_x" or simply "fuel NO_x." This "fuel NO_x" plus the "thermal NO_x" make up the total NO_x emitted from the stack. In some cases with a high nitrogen content fuel, the "fuel NO_x" can contribute more than half of the total NO_x emissions.

The formation of "fuel NO_x" is not presently well understood. There is some evidence that "fuel NO_x" can be reduced by lowering the available oxygen in the flame. Switching to a fuel which contains less nitrogen has also produced lower total NO_x; however, this is not always predictable for fuel variations within the same fuel type (i.e., #6 oils, bituminous coals, etc.). It should also be mentioned that measurements of total NO_x at the stack or furnace outlet will not distinguish between "thermal NO_x" and "fuel NO_x." NO_x variations measured under various boiler operating modes may result from combined effects on both NO_x components.

Finally, while oxygen availability and flame temperature appear to be the major combustion parameters affecting NO_x formation, the absolute levels of NO_x emitted from the stack will also depend on a variety of other factors.

Different types of fuels generally exhibit a different range of NO_x emissions. This is due to differences in fuel nitrogen already mentioned plus fundamental differences in the fuel burning characteristics, general requirements of the associated burning equipment (stoker versus oil gun) and so on. For a particular fuel composition, boiler-to-boiler variation in NO_x emissions also suggests the importance of burner and furnace design and the specific flow patterns of the flame within the furnace. At any given boiler with a constant fuel, the NO_x emissions are also generally dependent on firing rate.

Heat Losses

The term "boiler efficiency" as used here pertains to the thoroughness by which the boiler extracts the total available heat energy from the fuel. Some loss of heat from the boiler is unavoidable and as a consequence boiler efficiency is always less than 100%. Boiler efficiency losses arise from four major sources:

1) Heat carried away (out the stack) by the hot flue gases. This loss is usually called "dry flue gas"

loss. Hotter stack temperatures and larger quantities of excess air increase this loss.

2) Latent heat of water vapor present in the flue gas. (Water vapor results from the combustion of hydrogen contained in the fuel and from moisture present in the fuel and combustion air.) This loss arises from the fact that water vapor contains more heat energy than an equal weight of liquid water. The latent heat would be recovered only if the water vapor were allowed to condense out before the flue gases leave the boiler. Since this would be impractical in conventional boiler systems (due to corrosion problems and reduced plume rise), the latent heat is not considered recoverable.

3) Unburned fuel and products of incomplete combustion. These include solid combustibles in refuse, carbon carryover from the furnace, and all other solid and gaseous combustible material carried away in the flue gases. This loss is usually referred to as "combustible loss' or "unburned fuel loss." Improperly adjusted burning equipment or an inadequate quantity of excess air can lead to rapid increases in combustible losses.

4) Heat lost from the boiler jacket through its insulation. This is generally termed "radiation loss" and includes heat radiated to the boiler room and the heat picked up by the ambient air in contact with the boiler surfaces. The quantity of heat lost in this manner is fairly constant at different boiler firing rates and as a result becomes an increasingly higher percentage of the total heat losses at the lower firing rates. Deteriorated insulation and furnace wall refractory will also increase these losses at all loads.

Figures 2.2 and 2.3 illustrate how the various efficiency losses are affected by changes in excess O_2 and firing rate. These results are based on actual measurements at a 13,000 lb/hr boiler fired with natural gas but the general trends are also representative of many natural gas, oil and coal fired boilers. Of course, the actual values of the various efficiency losses will be different depending on specific fuel properties, boiler/burner design, operating conditions, etc.

FIGURE 2.2: VARIATION IN BOILER EFFICIENCY LOSSES WITH CHANGES IN EXCESS O_2

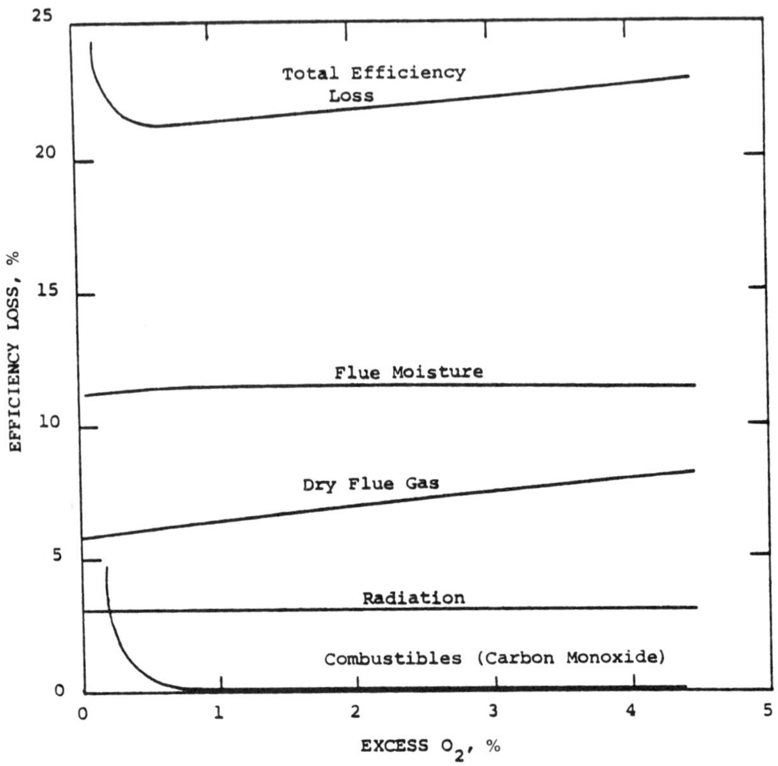

Source: PB-264 543

At each firing rate the highest boiler efficiency is achieved by firing with the lowest practical excess O_2. For the particular example in Figure 2.2, the point of maximum efficiency occurred at about 0.5% excess O_2 where the total efficiency losses were at a minimum. (Excess O_2 levels at peak efficiency can vary significantly from boiler to boiler and with different fuels.) A futher reduction in excess O_2 caused rapid increases in carbon monoxide and other combustibles which resulted in a very rapid deterioration in efficiency.

FIGURE 2.3: VARIATION IN BOILER EFFICIENCY LOSSES WITH CHANGES IN BOILER FIRING RATE

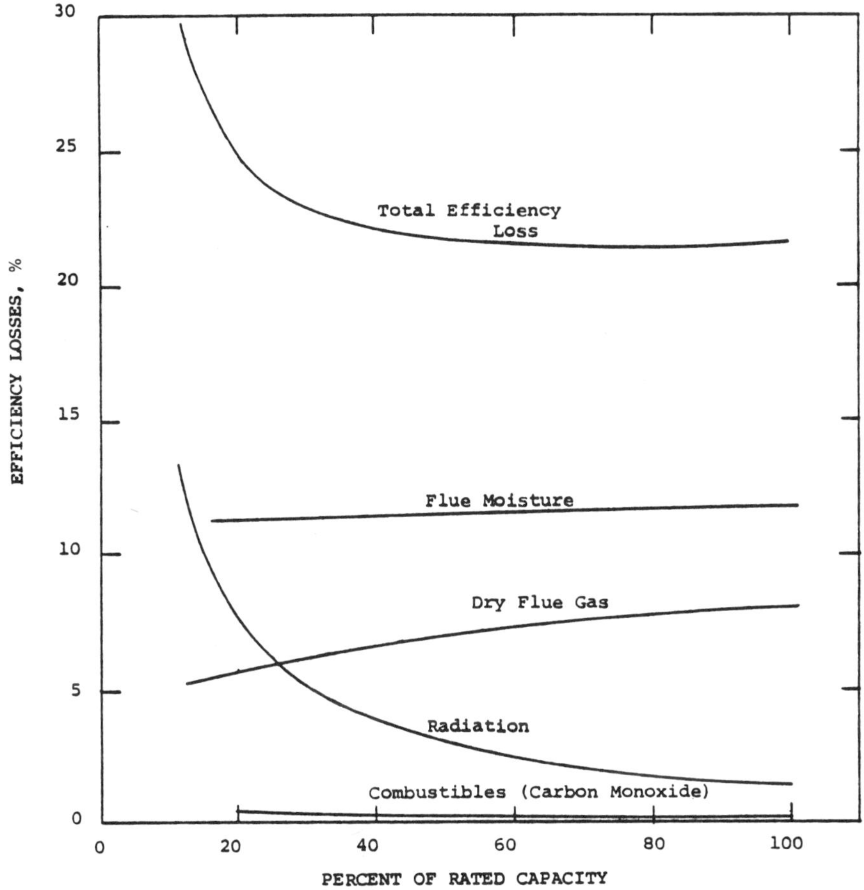

Source: PB-264 543

With high excess O_2, some fuel is used to heat unneeded excess air to the stack temperature where it is carried out the stack (resulting in higher "dry flue gas" losses and reduced efficiency). It might at first seem that large quantities of excess air would mix with the flame and cool it, resulting in a lower stack gas temperature. In practice, however, the reverse is true. High excess air does cool the flame and cool the furnace. But, at the same time, the boiler must fire more fuel to generate the same quantity of steam and proportionately greater amounts of air are fed through the burner. Consequently, greater amounts of gas pass through the boiler so quickly that they retain their heat instead of transferring it to the tubes. This explains why stack temperatures generally increase as the excess O_2 is raised, contributing to reduced efficiency.

It would be desirable to operate at the point of peak efficiency at each firing rate to minimize fuel wastage. However, there are certain limitations in typical boiler combustion controls which require that boilers be operated at some point slightly below peak efficiency to avoid combustibles and potentially unsafe conditions.

Figure 2.3 illustrates that there are significant changes in efficiency losses as the firing rate is varied. Especially obvious are the "radiation loss" which increases at lower firing rates and the "dry flue gas" loss which increases at higher firing rates. These characteristics and the "total efficiency loss" curve shown in the figure are very representative of many industrial boilers. Lowest losses (highest efficiencies) occur over the range of firing rate from approximately 50 to 80% of capacity. Operating the boiler in this range as much of the time as possible would therefore be desirable from a fuel efficiency standpoint.

Once the basic approach and test equipment necessary to identify the point of maximum efficiency have been determined for a particular boiler, this information can be used together with NO_x emissions measurements to determine the optimum boiler operating practice for high efficiency and reduced NO_x.

The efficiency improvement for a boiler at each firing rate will depend on the reductions in excess O_2 possible and the stack temperatures. To aid in determining the actual efficiency improvements, Figure 2.4 gives the approximate percent efficiency improvement corresponding to each 1% reduction in excess air.

Since reductions in excess O_2 will be measured directly, use Figure 2.1 to convert excess O_2 to excess air and then determine the efficiency improvement at the proper stack temperature. For example, if the excess O_2 was reduced from 6 to 3.5%, from Figure 2.1 it is determined that the excess air dropped from 40 to 20%, a 20% total change. If the stack temperature was 400°F, the efficiency improvement was 20 times 0.058 or 1.16%.

Reductions in stack temperature will also generally occur as the excess O_2 is lowered. To account for this effect, perform the previous calculation using the initial stack temperature and add to this improvement the efficiency improvement obtained from Figure 2.5. This additional improvement from Figure 2.5 is the effect of reduced temperature and is determined using the final lower excess air level. To illustrate, if in the previous example the stack temperature dropped from 400° to 370°F, as the excess O_2 was lowered from 6 to 3.5%, the effect due to the 30°F lower stack temperature would be 3 times 0.252 which equals 0.76%. The total improvement in efficiency is then 1.16 plus 0.76 for a total improvement of approximately 1.92%. Figures 2.4 and 2.5 can be used to estimate efficiency improvements achieved at constant firing rates on natural gas, #2 through #6 oils and common bituminous and sub-bituminous coals.

FIGURE 2.4: PERCENT EFFICIENCY IMPROVEMENT PER EVERY 1% REDUCTION IN EXCESS AIR

Valid for estimating efficiency improvements on typical natural gas, #2 through #6 oils and coal fuels.

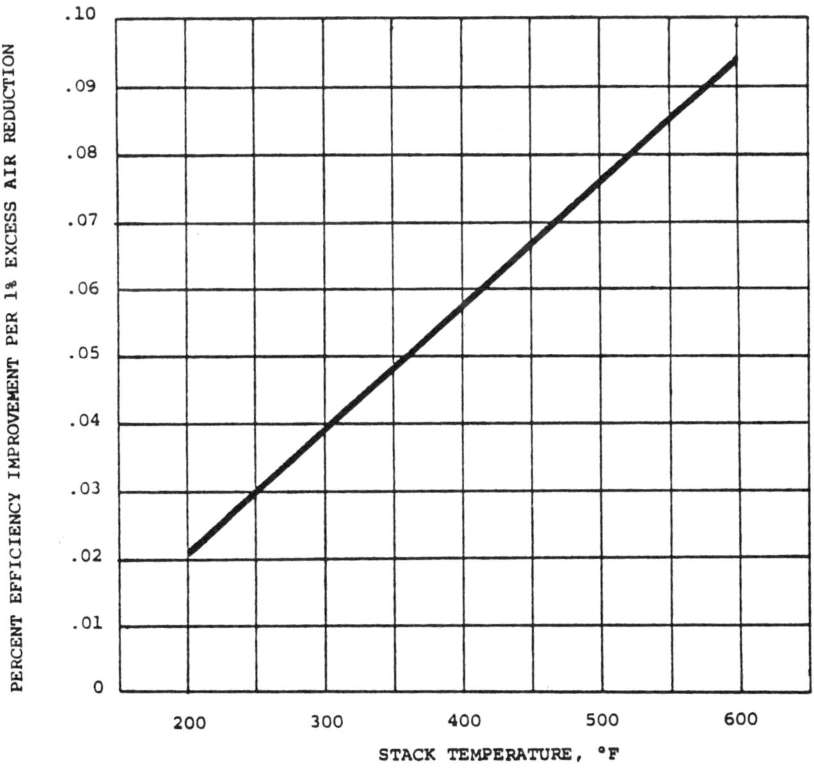

Source: PB-264 543

FIGURE 2.5: PERCENT EFFICIENCY IMPROVEMENT PER EVERY 10°F DROP IN STACK TEMPERATURE

Valid for estimating efficiency improvements on typical natural gas, #2 through #6 oils and coal fuels.

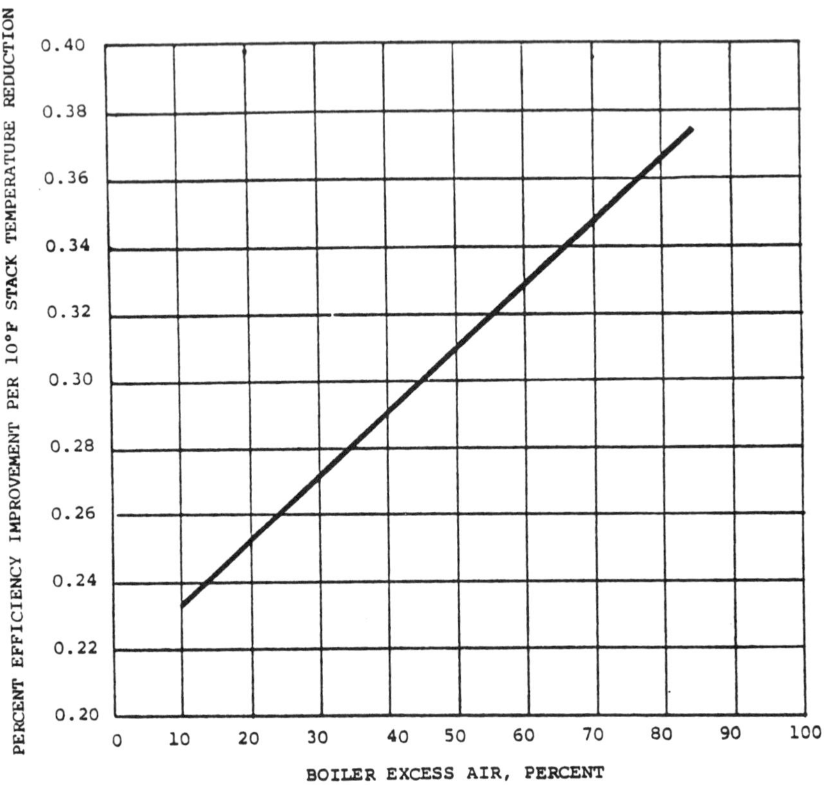

Source: PB-264 543

Some boiler operators may find that efficiency can be improved by 1 or 2%, which might at first seem to be a small improvement having little effect on total fuel costs. However, even for the smallest boilers, these seemingly small improvements can lead to significant annual fuel savings which easily justify the time and effort spent in achieving the improvements.

Consider as an example a 10,000 lb/hr boiler that is operated year-round with an average annual loading of one-half full capacity. If the efficiency is improved by 2% over the boiler operating range, the annual fuel consumption will decrease by 2.5%, assuming the boiler efficiency is approximately 80%. (Percent fuel reduction equals efficiency improvement times 100, divided by the efficiency.) If the boiler is fired with #6 oil costing $2.00 per million Btu, the annual fuel savings would be nearly $2,800. For larger boilers, this savings would increase in proportion to the size of the boiler ($28,000 for a 100,000 lb/hr boiler and so on).

EFFECT OF IMPORTANT PARAMETERS

Effect of Combustion Air on Efficiency

For the following study the combustion efficiency, η_C, is defined as the energy released in the combustion chamber divided by the energy contained in the higher heating value of the fuel. (The higher heating value can be considered to be the ultimate energy that could be obtained.) The combustion efficiency does not include heat transfer to the surroundings, blowdown, etc. However it will be nearly equal to the overall boiler efficiency since these latter effects are usually small. The change in overall efficiency will always be approximately the same as the change in combustion efficiency. Thus, the economic loss based on a change in combustion efficiency will always be the same as that based on overall boiler efficiency.

Figure 2.6 is a plot of efficiency versus percent excess air. For this case combustion is assumed to be complete. The major cause of decreasing efficiency with increasing excess air is the increased mass (and energy) which leaves as flue gas. The curves indicate that 200% theoretical air produces approximately a 3-8% decrease in efficiency, depending on flue gas temperature. Graphs are presented for methane and No. 2 fuel oil and each graph shows flue gas temperature as an independent parameter. The efficiency drops with increasing flue gas temperature.

Figure 2.7 is a plot of efficiency versus percent theoretical air for deficient air. In this case, incomplete combustion is the major cause of decreasing efficiency. The drop in efficiency is much more pronounced than with excess air, e.g., the use of 50% theoretical air, with methane, results in a drop in efficiency of more than 50%. This shows that it is essential to always maintain complete combustion. The curves for deficient air were weakly dependent on flue gas temperature because the energy loss associated with incomplete combustion is several orders more than the sensible energy loss in the flue gas.

FIGURE 2.6: EFFECT OF EXCESS AIR ON EFFICIENCY

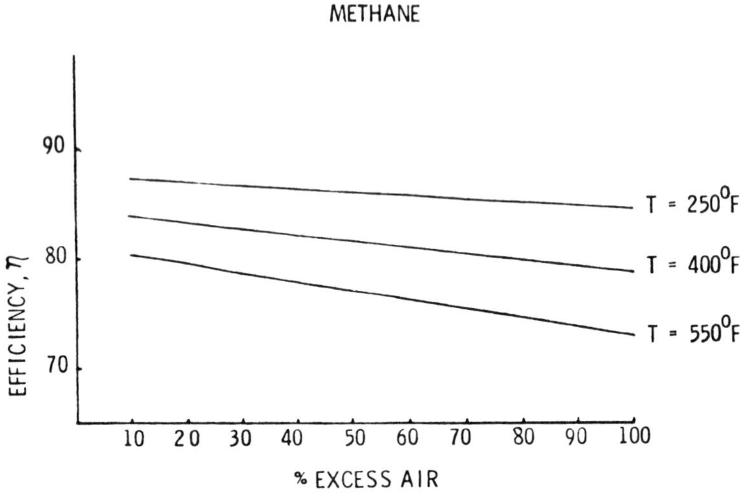

T = Flue Gas Temperature

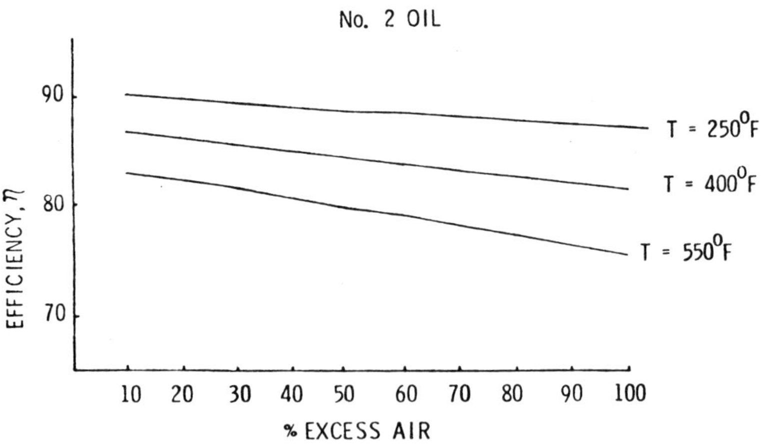

T = Flue Gas Temperature

Source: PB-265 713

FIGURE 2.7: EFFECT OF DEFICIENT AIR ON EFFICIENCY

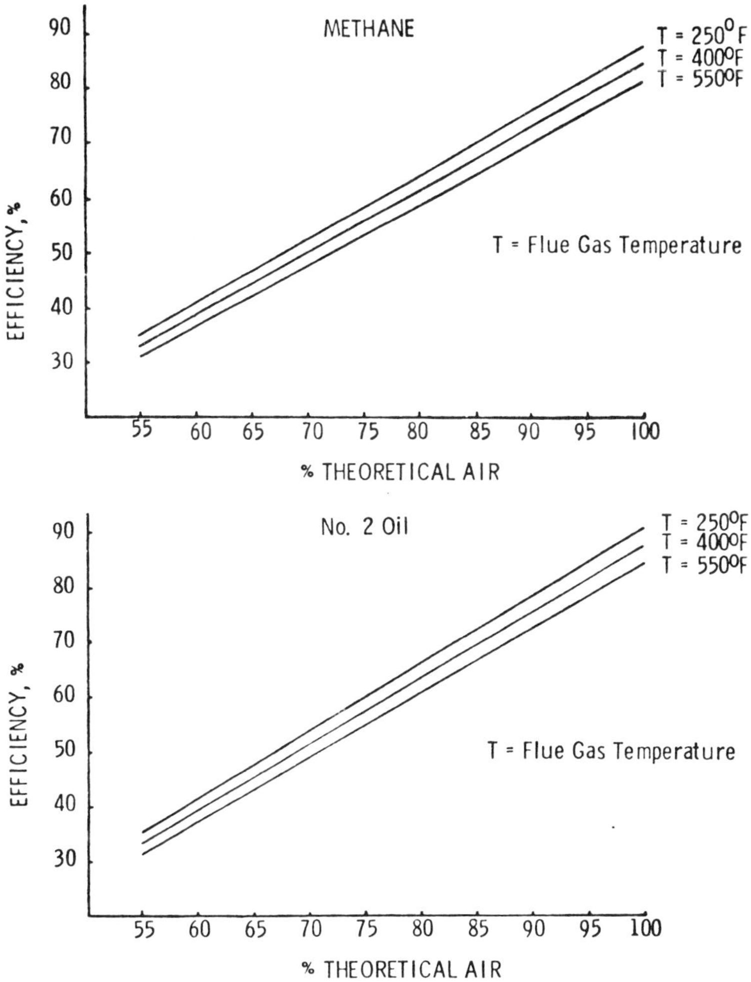

Source: PB-265 713

Figure 2.8 is a plot of efficiency versus ambient pressure with percent oxygen in the flue gas as a parameter. This figure was obtained assuming that the combustion system is designed to deliver a fixed volume of air. (Some boilers are provided with automatic controls to regulate the volume flow rate to achieve the proper air volume flow rate.) If the temperature and volume flow rate are fixed, pressure is directly proportional to the mass flow rate. The zero percent oxygen curve is used as a base line. The peak efficiency occurs at 14.7 psia, the design condition. If the pressure is greater than 14.7 psia then a condition of ex-

FIGURE 2.8: THE EFFECT OF AMBIENT AIR PRESSURE

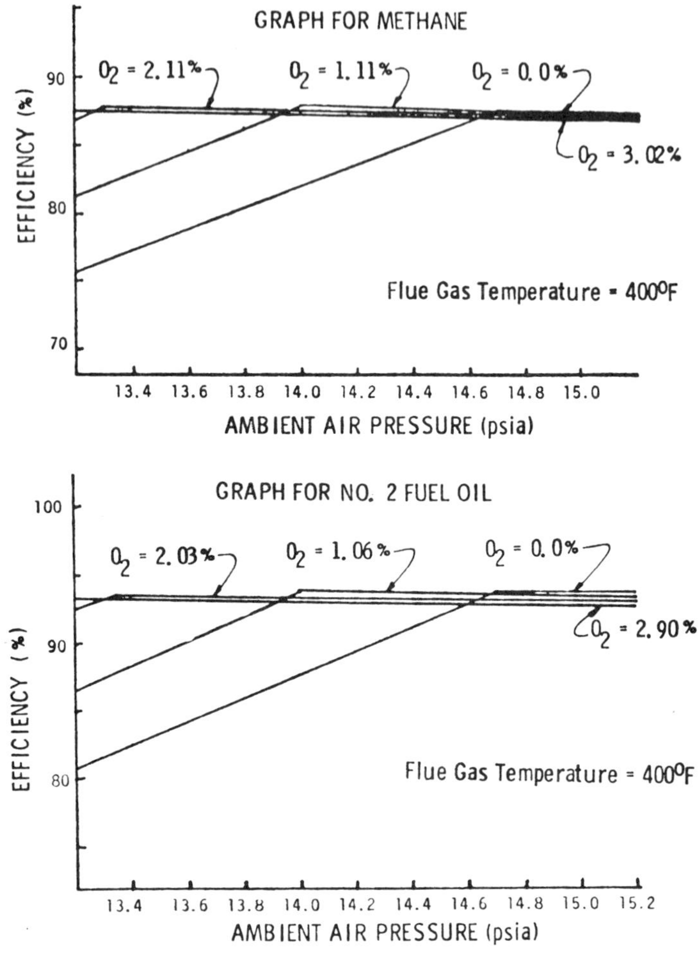

Source: PB-265 713

cess intake air exists. For pressure less than 14.7 psia insufficient air is provided and incomplete combustion occurs.

For small percentages of oxygen in the flue gas the curves show that there is a wide operating range in ambient pressure for which the efficiency is essentially constant. These curves show that a boiler can be operated with only a small amount of excess air without the necessity to frequently change the air supply for ambient pressure effects. Figure 2.8 gives results for methane and No. 2 fuel oil at a flue gas temperature of 400°F.

Figure 2.9 is a curve similar to Figure 2.8 in that ambient temperature replaces ambient pressure as the independent variable. With pressure and volume flow rate fixed the mass flow rate varies inversely with the temperature. The zero percent oxygen curve is used as a base line. The peak efficiency occurs at 77°F, the design condition. The drop in efficiency on either side of the design point is due to excess intake air or incomplete combustion as before. Again the curves with excess oxygen in the flue gases serve to extend the operating range of the boiler by avoiding incomplete combustion. These curves are also given for methane and No. 2 fuel oil at a flue gas temperature of 400°F.

FIGURE 2.9: THE EFFECT OF AMBIENT AIR TEMPERATURE

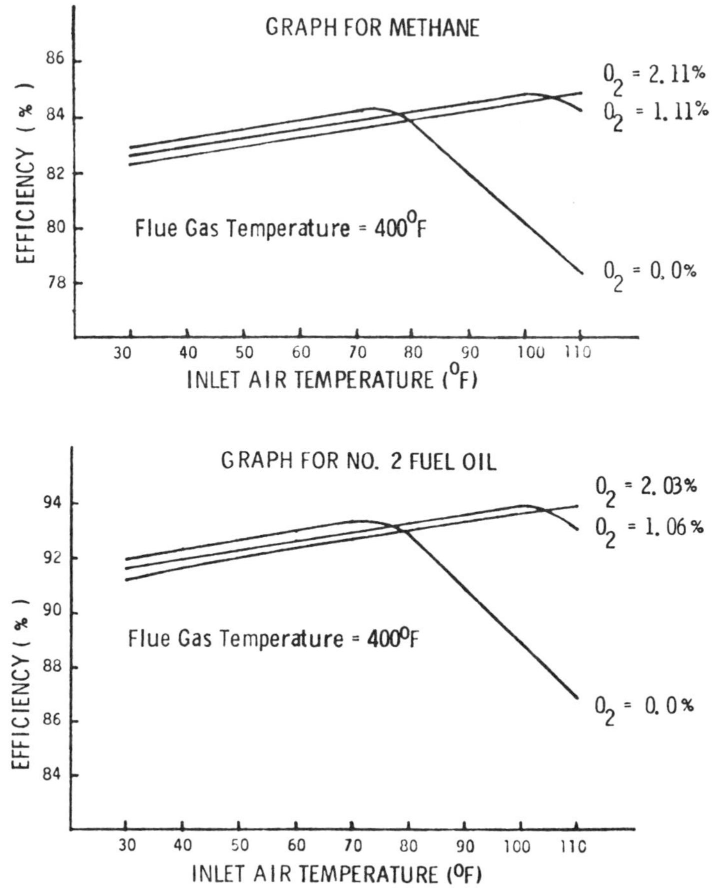

Source: PB-265 713

Figure 2.10 is a plot showing the effect of water vapor in the intake air. The curve for zero percent oxygen is the base line. The effect of water vapor is to reduce the amount of air coming into the boiler. The curves with excess oxygen show that the effect of the water vapor itself is negligible and that the reduced air flow tends to increase the efficiency. Curves are given for methane and No. 2 fuel oil at a flue gas temperature of 400°F.

FIGURE 2.10: THE EFFECT OF WATER VAPOR ON EFFICIENCY

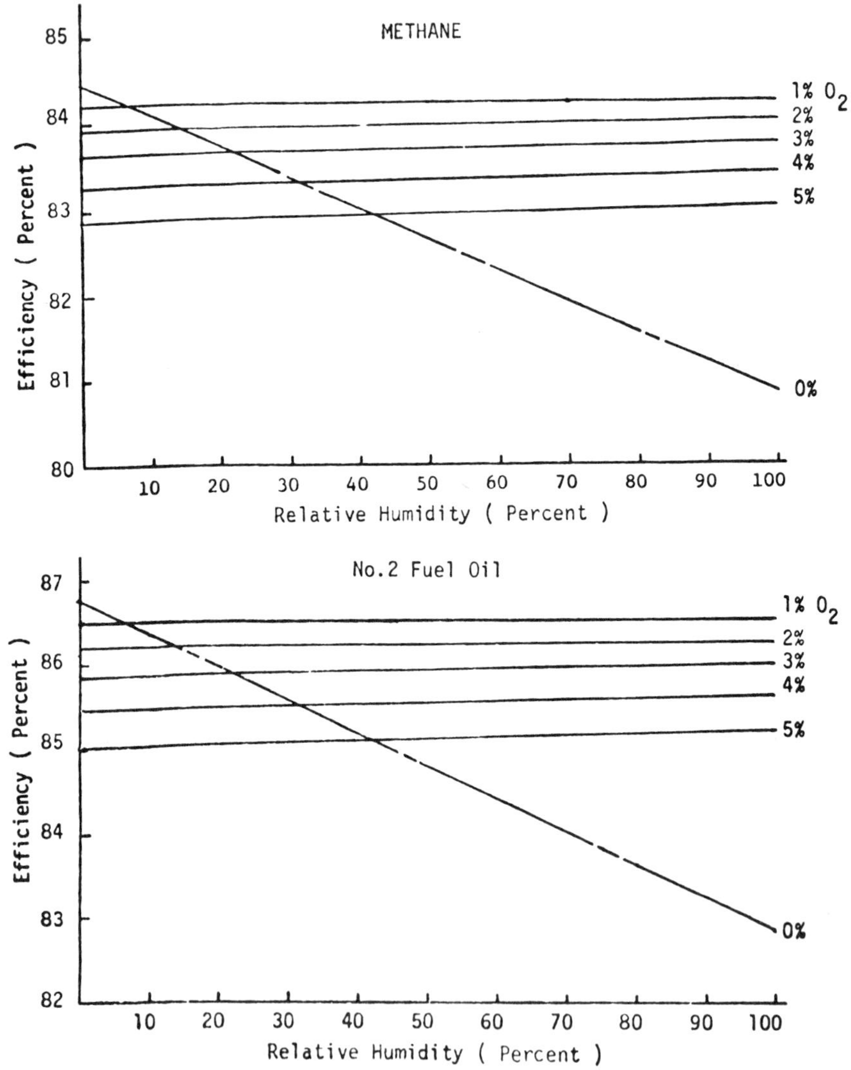

Source: 265 713

Effect of Temperature of Inlet Water on Efficiency

The efficiency of a boiler can be improved by using the flue gas energy to preheat the inlet water. The combined system is shown in Figure 2.11.

FIGURE 2.11: BOILER SYSTEM WITH WATER PREHEAT

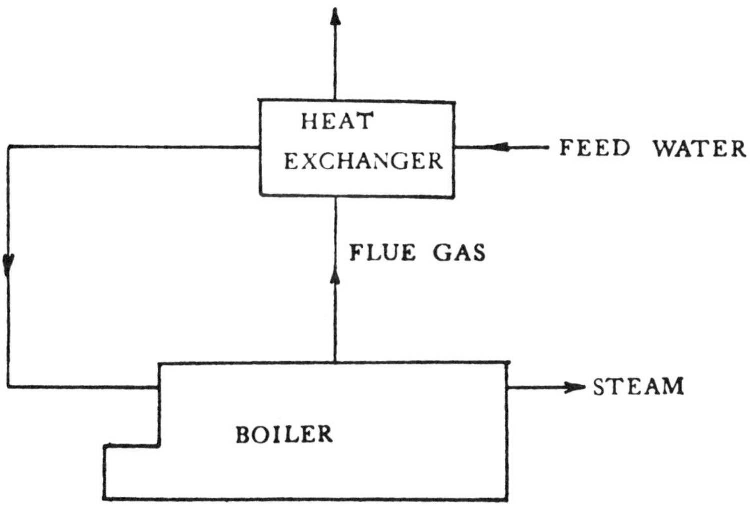

Source: PB-265 713

A typical magnitude for improved efficiency can be obtained by considering a practical example as follows. Feedwater is available at 70°F and the boiler is used to produce saturated steam at 350°F. The efficiency of this system compared to that of a boiler without feedwater preheat is shown in Figure 2.12. This curve shows that feedwater preheat can improve efficiency by as much as 10%. In general if the flue gas temperature is high (e.g., 500°F or greater), than a condition of insufficient heat transfer surface exists and the addition of a heat exchanger to preheat the feedwater can be highly desirable.

For well designed and maintained boilers, the flue gas temperature will be near the minimum value allowed to prevent corrosion. Feedwater preheat will not pay out under this condition. It is always undesirable to preheat the water with any energy source which is not free. It can be shown that in this case the boiler efficiency remains essentially constant while the amount of steam produced decreases. The net results of this practice would be a boiler system operating at essentially the same fuel efficiency but with decreased throughput.

FIGURE 2.12: EFFECT OF FEEDWATER PREHEAT ON BOILER SYSTEM EFFICIENCY

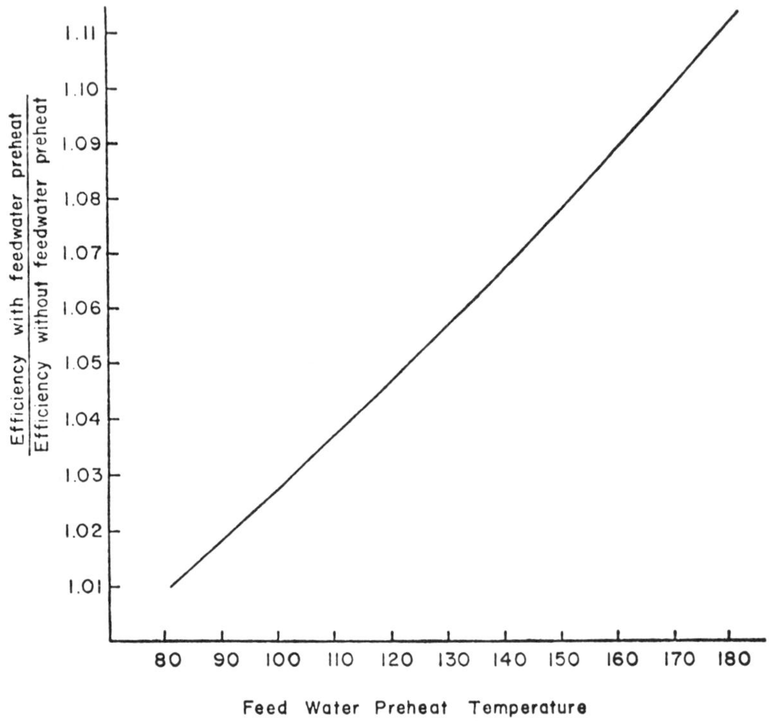

Source: PB-265 713

Effect of Flue Gas Temperature on Efficiency

Figure 2.13 shows efficiency against flue gas temperature for various constant amounts of theoretical air. Note that the drop in efficiency can be very significant particularly if the system is operating with excessive amounts of air. This occurs because increasing the air flow also increases the mass and thus energy discharge in the flue.

FIGURE 2.13: EFFECT OF FLUE GAS TEMPERATURE ON EFFICIENCY

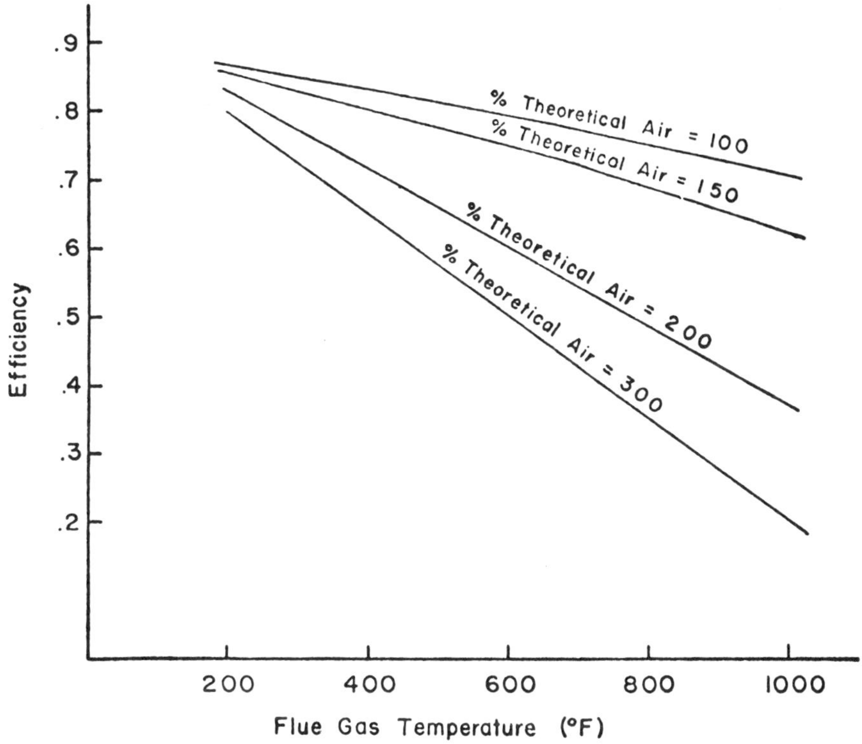

Source: PB-265 713

Effect of Scale on Boiler Surfaces

Any scale formation provides an additional "resistance" to heat transfer over the case of a clean surface. Thus, the heat transfer through scale can be written

$$q = k\frac{\Delta T}{L}$$

where L is the scale thickness, k is its thermal conductivity and ΔT is the temperature drop. Table 2.1 (1) gives the thermal conductivity of various scale deposits as well as steel. This table shows that k for scale is more than an order of magnitude lower then the value for steel.

Thus, any scale formation has a tremendous potential to increase the resistance and thus decrease the heat transfer. Figure 2.14 shows the reduction in heat transfer versus deposit thickness for various types of deposits.

TABLE 2.1: THERMAL CONDUCTIVITY OF SUBSTANCES FOUND IN BOILERS

Material	Thermal Conductivity*
Analcite	8.8
Calcium Phosphate	25
Calcium Sulfate	16
Magnesium Phosphate	15
Magnetic Iron Oxide	20
Silicate Scale	0.6
Fire Brick	7
Insulating Brick	0.7
Boiler Steel	310

*Btu/(ft^2 hr °F in^{-1})

Source: PB-265 713

FIGURE 2.14: EFFECT OF SLUDGE ON BOILER HEAT TRANSFER

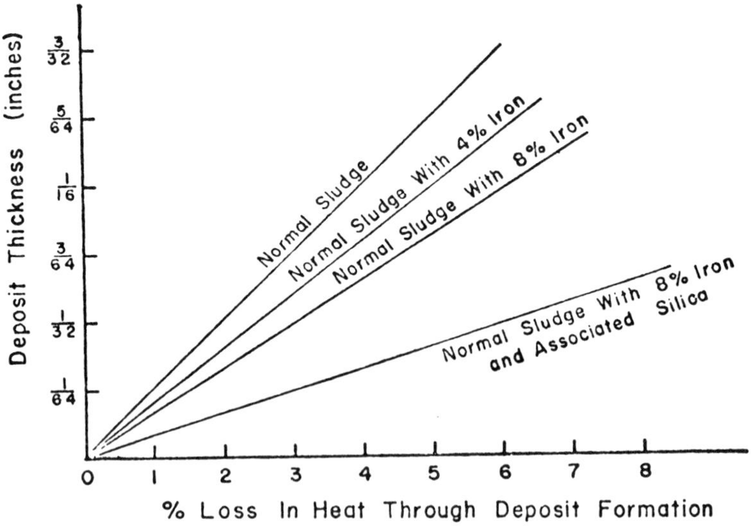

Source: PB-265 713

Effect of Fuel Composition on Efficiency

On boilers with combustion systems which are designed for a fixed air intake based on a particular fuel the effect of varying the chemical composition of the fuel can be significant. To illustrate this point Figure 2.15 is presented, which shows efficiency versus the ratio of hydrogen atoms to carbon atoms in the fuel. Figure 2.15 shows curves for both methane and No. 2 fuel oil boilers. Consider the curve for the boiler burning methane. The efficiency is a peak at the design condition (ratio of hydrogen to carbon of four). To the right of this point there is an excess of hydrogen and a rapid drop in efficiency. This corresponds to a condition of deficient air which results in incomplete combustion. The strong effect on incomplete combustion has already been noted.

FIGURE 2.15: EFFECT OF FUEL COMPOSITION ON EFFICIENCY

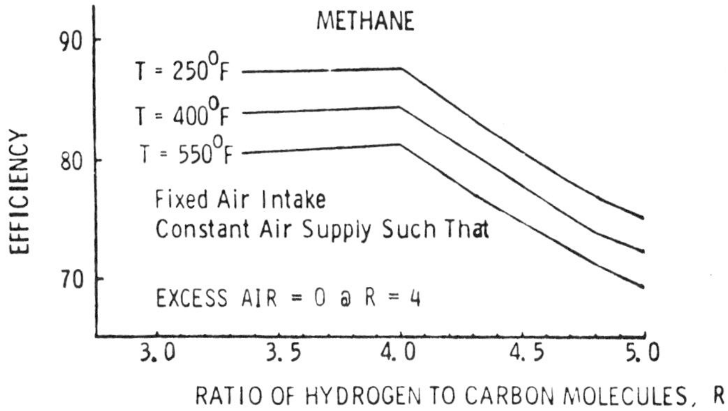

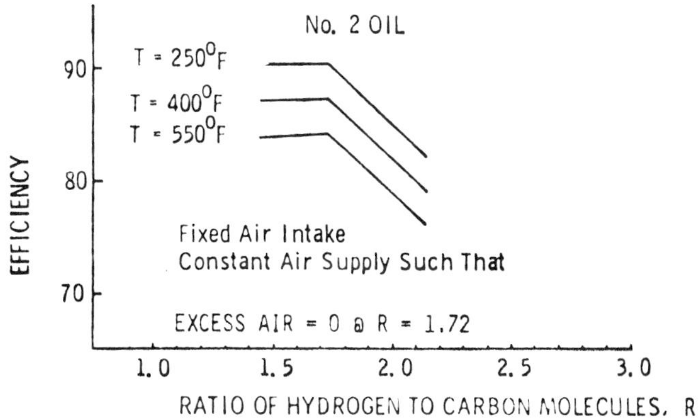

Source: PB-265 713

For a hydrogen content less than the design condition, the air will be in excess of that required giving the slow decrease in efficiency described earlier in the case of excess air. The curves for No. 2 oil follow the same general trends (note that the design hydrogen to carbon atom ratio is 1.72). These curves were obtained by varying the ratio of hydrogen to carbon atoms in the fuel but not the chemical heating value of the fuel.

Reference

(1) Phelan, J.V. and L.R. Gelosa, "How to Control Boiler Iron Deposits," *Chemical Engineering,* Vol. 82, No. 5, March 1975, pp. 174-178.

AIR POLLUTANTS GENERATED BY COMBUSTION

The exhaust gases from all combustion devices contain a variety of byproducts, some of which are considered environmental pollutants. Some of these pollutants are immediately toxic in themselves while others pose indirect health hazards by virtue of their ability to react with other pollutants in the air to form more hazardous compounds. Furthermore, these pollutants also differ by the way they originate in the combustion processes. Some are more related to the composition of the fuel itself whereas others are dependent on the characteristics of the actual fuel burning process and as such are sensitive to the design and operating variables of the combustion device.

The EPA has identified six major pollutants and established emission standards to protect the public health and welfare. These six "criteria" pollutants are: Particulate matter, sulfur dioxide, nitrogen oxides, carbon monoxide, hydrocarbons, and oxidants.

These substances are described here, including the principal health effects, their formation in boiler combustion systems and methods currently used to limit their emissions from boiler stacks. This information should help familiarize boiler operators and others with the major aspects of combustion pollutants which have prompted action by the EPA and some state and local air quality agencies to limit these emissions from many sources.

Particulate Matter

Particulate matter is the nongaseous portion of the combustion exhaust, consisting of all solid and liquid material (except water droplets) suspended in the exhaust gases. They can be generally defined as any material that would not pass through a very fine filter. Particulates originating in the combustion process can range in size from submicron in diameter (less than one millionth of a meter) to diameters larger than a millimeter (thousandth of a meter). The larger particulates do not carry far in the atmosphere and usually fall to the ground near the source. The small particles, which make up the bulk of particulate matter, can remain in the atmosphere for long periods of time and contribute to haze and obscured long-range visibility. These "fine particles" are potentially the most hazardous to health since they are easily carried into the small passages of the respiratory tract during normal breathing.

Particulate matter can be composed of a wide variety of materials including unburned fuel, sulfur compounds, carbon, ash constituents in the fuel (including many toxic metals), and even noncombustible airborne dust that enters the combustion system with the combustion air. Many of these materials by themselves are identified as health hazards and when dispersed in the air in fine particles can be inhaled and subsequently absorbed into the body. High concentrations of particulate matter in the inhaled air can have more direct health effects by irritating or blocking the surfaces of the respiratory tract leading to temporary or permanent breathing impairments and physical damage. Oral ingestion of particulate matter is also possible from particulate fall-out onto vegetation and food crops.

The quantity and composition of particulates generated in boilers are influenced by several factors including the type of fuel being burned, the boiler operating mode, and combustion characteristics of the burners and furnace. Burner designs and operating modes that tend to promote thorough, efficient combustion generally will reduce the fraction of combustible material in the particulates. From a boiler operating standpoint, it is desired to minimize these fuel-derived materials since they represent wasted available fuel energy and can lead to troublesome internal furnace deposits or objectionable smoke at the stack. Burner air/fuel ratio is one of the more important operating parameters that can influence the quantity of combustible particulate material generated.

Natural gas and most oil fuels are referred to as "clean burning" fuels primarily due to their lower tendency to form solid combustibles (smoke, soot, carbon, etc.) and their low ash content. By comparison, some heavy oils and most coal fuels contain substantial quantities of ash which subsequently form the bulk of the noncombustible particulate matter generated in the furnace. For the case of coal, ash can compose up to 20% or more of the total weight of fuel burned. Preventing its accumulation on internal boiler surfaces is a major consideration in the design of the boiler.

In some boiler designs, most of the coal ash remains in the exhaust gases leaving the boiler, but the use of various particulate control devices allows low concentrations to be emitted from the stack. Current devices employ various techniques to remove particulates from the stack gases. These include filtration, mechanical separation and electrostatic precipitation. Many designs have proved to be capable of removing more than 99% of the particulates and are also applicable to oil fired boilers where particulate controls are required.

Sulfur Dioxide

SO_2 is a nonflammable, colorless gas that can be "tasted" in concentrations of less than 1 part per million in the air. In higher concentrations, it has a pungent, irritating odor.

Sulfur dioxide (SO_2) is formed during the combustion process when sulfur (S) contained in the fuel combines with oxygen (O_2) from the combustion air. Sulfur trioxide (SO_3) is another oxide of sulfur which is also formed in this manner. SO_2 together with SO_3 comprise the total oxides of sulfur, generally referred to as "SO_x." SO_3 is usually no more than 3 to 5% of the total SO_x generated in the boiler.

Except for sulfur compounds present in particulate matter, all of the sulfur initially contained in the fuel is converted to SO_2 and SO_3. Before leaving the stack, SO_3 can combine with moisture in the exhaust gases to form sulfuric acid which condenses onto particulates or remains suspended in the stack gases in the form of an acid mist. In the atmosphere, a portion of the SO_2 is converted to SO_3 which similarly forms sulfuric acid by combining with moisture in the air. The SO_3 can also form other sulfur compounds such as sulfates.

The sulfates and acid mists can contribute significantly to reduced visibility in the atmosphere. Corrosion of materials exposed to the air and damage to vegetation are other major environmental effects. The health effects of sulfur oxides, sulfuric acids and some of the sulfates are primarily related to irritation of the respiratory tract. These effects may be temporary or permanent and include constriction of lung passages and damage to lung surfaces.

The quantity of SO_x generated in the boiler is primarily dependent on the amount of sulfur in the fuel and is not highly affected by boiler operating conditions or design. Regulating the quantity of sulfur allowed in the fuel is a primary method of controlling SO_3 emissions. Stack gas "scrubbers" which remove SO_2 from the combustion exhaust gases can also be effective where "high sulfur" fuels are used.

Nitrogen Oxides

Nitric oxide (NO) and nitrogen dioxide (NO_2) are the two forms of nitrogen oxides generated by boiler combustion processes. Together, these compounds are customarily referred to as total oxides of nitrogen or simply "NO_x." NO is a colorless, odorless gas and is not considered a direct threat to health at concentrations found in the atmosphere. NO_2 is a considerably more harmful substance. Although NO_2 comprises typically 5% or less of the NO_x emitted from boiler stacks, a large fraction of the NO is converted to NO_2 in the atmosphere.

NO_2 is a yellow-brown colored gas which can affect atmospheric visibility. It also has a pungent, sweetish odor that can be detected at concentrations sometimes reached in polluted air. In much higher concentrations (100 ppm) NO_2 can be fatal when inhaled. Prolonged exposures at much lower concentrations can cause cumulative lung damage and respiratory disease.

NO_x is formed spontaneously during the combustion process when oxygen and nitrogen are present at high temperatures. All three ingredients (oxygen, nitrogen and high temperature) are essential elements of the combustion process and it would therefore be very difficult to prevent the formation of NO_x altogether. Nitrogen is present in the combustion air and in the fuel itself. Minimizing the fuel nitrogen content has been shown to reduce NO_x but this is not currently a practical NO_x control approach. Most NO_x reduction techniques currently applied to boilers are effective as a result of lower peak flame temperatures in the furnace, reduced availability of oxygen in the flame or a combination of both. These techniques include low excess air operation, fuel-rich (staged) firing and flue gas recirculation. While these approaches limit the formation of NO_x in the furnace, future techniques may be developed that "scrub" the NO_x from the exhaust gases before entering the stack. Fuel desulfurization processes and other fuel "cleaning" treatments may have some associated benefits in reduced fuel nitrogen content. For a more detailed description of NO_x removal, see Chapter 5.

Carbon Monoxide

Carbon monoxide (CO) is a product of incomplete combustion and its concentration in boiler exhaust gas is usually sensitive to boiler operating conditions. For example, improper burner settings, deteriorated burner parts and insufficient air for combustion can lead to high CO emissions. CO measurements at the stack are often used as an indicator of poor combustion conditions.

CO is an invisible, odorless, tasteless gas. Exposure to CO-containing exhaust gases produces a well known "CO poisoning" which can be fatal. CO emitted from boiler stacks is dispersed in the atmosphere and together with CO from other sources is generally not in high enough concentrations to produce any immediate health effects (an exception might be in the vicinity of high density automobile traffic). However, there is concern that long-term exposures to these concentrations may cause eventual health problems.

Hydrocarbons

Like carbon monoxide, hydrocarbons are indicative of incomplete or inefficient combustion and can be essentially eliminated from the boiler stack gases by proper operation of the fuel burning equipment. However, this may be misleading since, strictly speaking, hydrocarbons cannot be entirely eliminated and trace quantities of hydrocarbon compounds will nearly always be present, regardless of how the boiler is operated.

Due to the tremendous variety of hydrocarbon compounds involved and the unknown health effects of some of these even in much larger quantities, it is difficult to assess their environmental impact. Some of these hydrocarbons resemble actual components in the fuel and are rightfully called "unburned fuel," while others are entirely modified forms generated in complex chemical reactions during the combustion process.

It is known that hydrocarbon air pollutants are important ingredients in the formation of photochemical smog. Under certain atmospheric conditions, they can also be transformed into other derivatives which are potentially more hazardous. Some of the manifestations of smog such as irritation of the eyes and respiratory tract are in part directly associated with hydrocarbons and their derivatives.

Oxidants

The term "oxidant" is generally applied to oxygen-bearing substances that take part in complex chemical reactions in polluted atmospheres. These so-called photochemical reactions, which are often intensified in the presence of sunlight, involve nitrogen oxides and reactive organic substances (including hydrocarbons and their derivatives) as the principal chemical ingredients. These react to form new compounds including ozone and PAN (peroxyacyl nitrates) which are usually considered to be the major oxidants in photochemical air pollution (smog). The term oxidant is also sometimes used to include the original chemical reactants, NO_x, hydrocarbons, and others.

While ozone and PAN are not generated directly in the boiler, the principal ingredients (especially NO_x) are supplied in part by exhaust gases from the boiler. By reducing the emission of these compounds from boilers and all the other combustion sources (automobiles, airplanes, etc.), their photochemical by-products, the oxidants, will be reduced.

Photochemical oxidants produce adverse effects on vegetable matter which can affect growth, and the quantity and quality of agricultural yields and other plant products. Deterioration of various materials (especially rubber) is also a well-known occurrence in polluted atmosphere which is attributed mainly to the presence of ozone.

A major effect on humans is irritation of the eyes. In quantities higher than typically found in polluted atmospheres, oxidants have an irritating effect on the respiratory tract producing coughing and choking. Headache and severe fatigue may be other side-effects. In lower concentrations found in polluted air, the effects appear less well defined. Aggravation of existing respiratory ailments such as asthma has sometimes been attributed to oxidants.

Combustion in Natural Gas and Fuel Oil Burners

The material in this chapter is based on a Federal Power Commission report (TID-27600).

Combustion takes place when fuel and air are brought together and an ignition source is provided.

To maximize efficiency in the furnace of a boiler, the following factors are important: Excess air in the combustion chamber should be held to the minimum consistent with complete combustion; the ash should contain no unburned combustible material; and the exit flue gas should contain no unburned combustible gases, such as carbon monoxide.

In industrial burner equipment, ignition is provided in the burner, and both fuel and air must be brought to this point. Fuels are always piped to the burner. In some installations air is piped under pressure to the burner, and in others the surrounding air is drawn into the burner, and no air piping is required.

Burners are installed on processing equipment of a variety of designs. The process requirements are, however, very common: Provide process heat in the amount required by the process and at the temperature required by the process.

The actual furnace, oven, or equipment design normally determines a relationship between process heat input and temperature. This allows, then, the heat input to the furnace to be controlled by controlling the temperature of the furnace.

In addition to temperature control, almost all furnaces have an additional control that maintans fuel and air flows at a constant preset value.

Maintaining this ratio between fuel and air is referred to as ratio control and this ratio can be adjusted to the proper level.

There are four major different burner types, each having applicable ratio-adjustment procedures. These different control systems have evolved over the years, and the application of any particular system has been governed by both personal preference and technical reasoning.

Each of the systems can be adjusted at an optimum A/F ratio and savings in fuel consumption after such adjustment range are characteristically between 10 and 30%.

TYPES OF BURNER SYSTEMS AND CONTROLS

Atmospheric Burners

The atmospheric burner is the simplest type of industrial burner. It consists of an air-to-gas mixer connected to a burner head. The burner head is made in various shapes to adjust the flame pattern to particular applications. Some common types are ring burner, pipe burner, torch burner and tunnel burner.

The ring burner is similar to that installed in home cooking stoves. The burner consists of a circular manifold with a connection that admits the air-gas mixture. A series of holes arranged around the top of the burner distributes the flame in a circular pattern.

The pipe burner is simply a section of pipe with a mixer installed at one end and a series of holes drilled along the length of the pipe to form a straight-line flame pattern.

The torch burner is a small burner tip that fits on the end of a mixer and produces a sharp-pointed flame.

The tunnel burner is similar to the torch burner but is fitted with a refractory block so that it can be installed in a furnace wall. It normally has a secondary air shutter to adjust the flow of air around the burner.

A venturi mixer is used to supply the proper air/fuel mixture to all the burners listed above. The venturi mixer is composed of a venturi tube, an air shutter, and a gas orifice (Figure 3.1). The gas orifice controls the gas flow, the venturi tube provides the proper aerodynamic shape for the entrainment of air and the mixing of air and gas, and the shutter controls the amount of air entrained into the mixture.

FIGURE 3.1: VENTURI MIXER

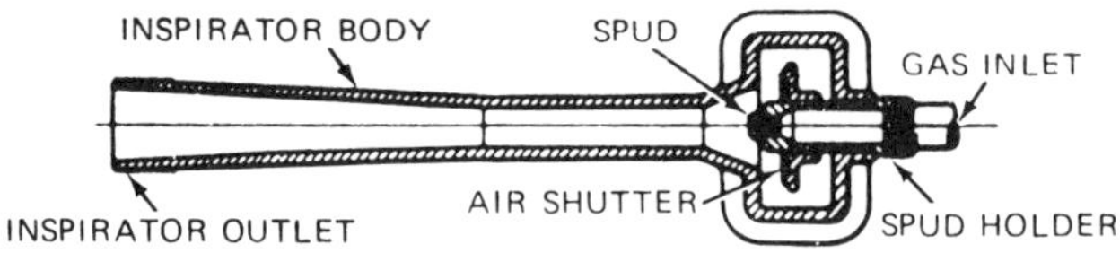

Source: TID-27600

In operation the gas stream issuing from the gas orifice entrains air from the surroundings. These air and gas streams mix in the venturi and form what is termed a partial premix. The desired quality of this partial premix varies from application to application in the range of 20 to 60% of the ideal or full premix condition. This quality is adjusted by either opening or closing the air shutter. Opening the shutter increases this quality, and closing the shutter decreases the quality. The effect of quality on burner operation is to lengthen or shorten the flame. When operating at a low-quality setting, the burner flame will be long, yellow, and soft or lazy in its appearance. When operating at a high-quality setting, the burner flame will be short, blue, and appear hard or sharp. The exact quality setting varies, depending on the heating operation being performed, and adjustments to this quality must be determined from past experience in operating the furnace equipment or from information supplied by the furnace equipment manufacturer.

As mentioned, the amount of air drawn into the venturi mixer is only a portion of the total air required to complete the combustion reaction. The balance of the air is supplied around the burner head. This air is commonly referred to as secondary air, and the adjustment of the flow of this air stream is critical to obtaining optimum combustion efficiency. The amount of secondary air admitted around the burner head is determined by the size of the opening around the burner head and by the amount of suction or draft in the furnace chamber. If the opening or draft is improper, the burner will operate "off ratio" and thus waste heat.

Typical Combustion Control Systems: The rate of fuel input to an atmospheric burner is controlled by a valve in the gas line to the burners (Figure 3.2). This valve can be either manually or automatically operated to control the temperature of the furnace. In either case the valve controls only the flow of gas. The primary air flow will roughly increase or decrease with the gas flow. The secondary air will not be greatly affected by the gas flow rate; it is controlled by the furnace draft or suction. The percentage oxygen in the flue gas therefore will increase with a decrease of the fuel input.

When this occurs, the efficiency of the furnace will decrease because the furnace is drawing in extra air that it is not using for combustion, and this air cools the furnace. For this reason a furnace with an atmospheric burner therefore should be set up to operate at maximum input approximately 80 to 90% of the time under normal load conditions. This is done by reducing the outlet pressure of the reducing regulator (Figure 3.2) until the percentage of time that the burner is on maximum fire is in the desired 80 to 90% range.

FIGURE 3.2: ATMOSPHERIC BURNER

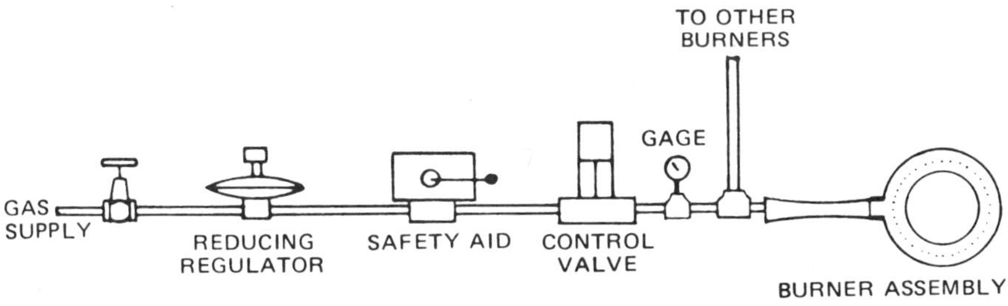

Source: TID-27600

Premix Burners

The premix burner is one of the most widely used industrial burners today. It is similar to the atmospheric burner in that its major components consist of an air-gas mixer and a burner head, but it differs in that a completely premixed air-gas mixture is delivered to the burner head and no additional air is required for proper combustion. This feature allows the premix burner to be operated in a sealed furnace and thus avoids air leakage into the furnace, which causes a decrease in furnace efficiency.

The two major components of the premix burner system are the air-gas mixer and the burner head. The air-gas mixer can be manifolded to several burner heads, as shown in Figure 3.3. Air-gas mixers fall into three major types: low-pressure aspirators, high-pressure inspirators and mechanical mixers.

Low-pressure aspirators consist of a venturi-shaped body with an air connection at one end and a mixture connection at the other. A gas connection with a gas orifice or an adjustment valve is located approximately in the center. The gas orifice or adjustment valve opening is changed by either drilling the orifice or adjusting the valve. These changes adjust the air/gas ratio for proper combustion. A gas passage in the throat area of the venturi admits gas to the air stream. The fuel gas is supplied at approximately 6 in. water column (this may vary for particular installations) to an atmospheric (or zero) regulator. This device automatically maintains the gas pressure to the inspirator at atmospheric pressure.

Air is supplied by a combustion air blower or fan to the aspirator at approximately 16 oz. pressure at the high-fire position. As the air passes through the venturi body, it creates a suction at the throat section, which pulls the gas through the gas orifice or adjustment valve and through the gas ports in the throat of the inspirator. This gas is then mixed with the air as they both pass through the last section of the venturi and on to the burner heads.

High-pressure inspirators are very similar in operation and appearance to atmospheric mixers. The major difference is that the fuel gas is supplied at a considerably higher pressure, usually in the range of 10 to 30 psi. This higher pressure level results in a higher velocity gas stream from the gas nozzle and thus enables the entrainment of all the air necessary for complete combustion. As in the atmospheric venturi mixer, an air shutter is used to control the A/F ratio of the mixture.

Mechanical mixers are usually used to feed premixed air and gas to one complete furnace or several furnaces. These centralized systems incorporate a compressor or blower that draws air and gas from a blender or carburetor and mixes them in the blower. The air-gas mixture is then manifolded to the various gas furnaces and/or burner heads. Mechanical mixers use a wide variety of carbureting systems to achieve control of the air/gas ratio.

As mentioned before, one mixer can be manifolded to one or many burner heads. The function of the burner head is to hold the flame in place and to shape the flame to a particular application. Burner heads fall into three general types: tip burner, line burner, and tunnel burner or cup burner.

The tip burner (also referred to as a blast tip, flame-retention tip, or torch tip) is a specially designed nozzle that produces a well-defined flame shape. There are numerous premix burner tips that can be designed. This nozzle can be fitted to a manifold along with many other nozzles or it can be attached directly to the air-gas mixer.

FIGURE 3.3: PREMIX PIPING SYSTEM

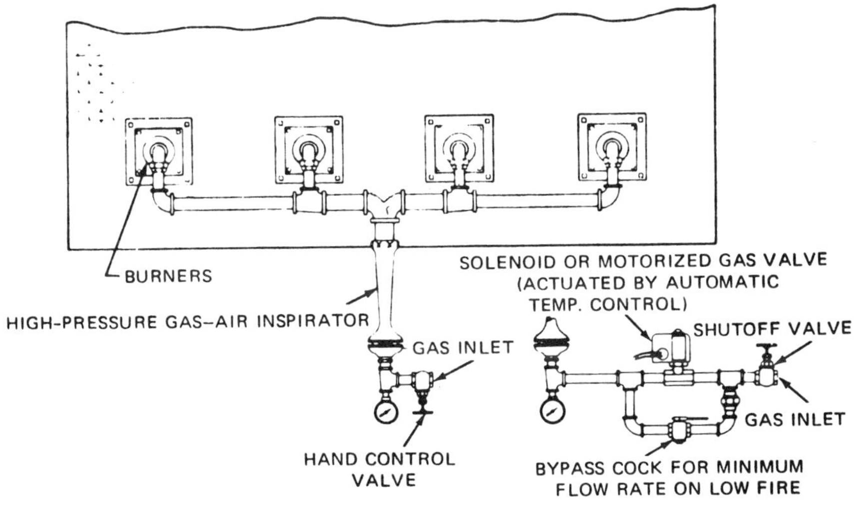

Source: TID-27600

The line burner is a straight assembly with holes drilled along its length and small flame piloting holes along each side of these main holes (Figure 3.4). This burner produces a series of small, sharp, blue flames along the length of the burner. Usually one air-gas mixer is used for each line burner.

The tunnel burner is simply a burner pot fitted with a refractory block that is mounted in the furnace casing (Figure 3.5). The burner produces an axial flame at stoichiometric ratio.

FIGURE 3.4: PREMIX LINE BURNER

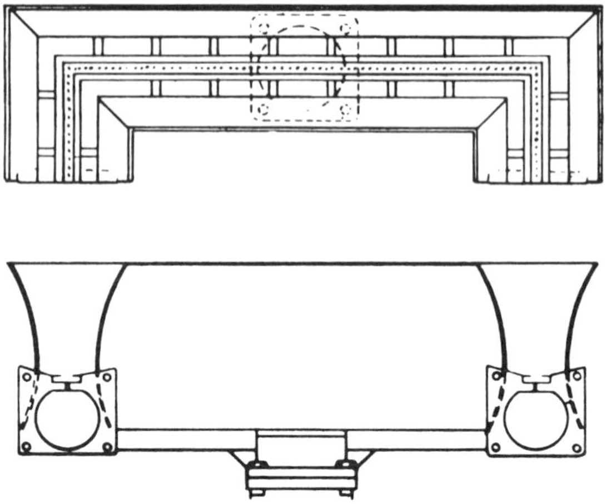

Source: TID-27600

FIGURE 3.5: PREMIX TUNNEL BURNER

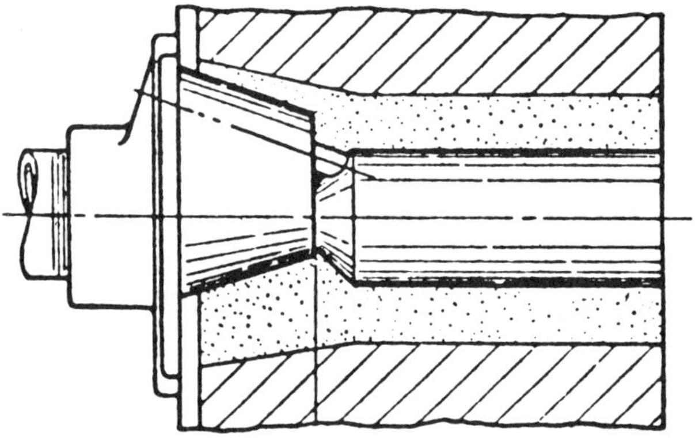

Source: TID-27600

Typical Combustion Control Systems: The firing rate of premix burners is controlled by varying the flow rate at which the air-gas mixture is delivered to the burners. During this variation of firing rate, the air/gas ratio of the mixture is maintained constant by the mixers.

On a low-pressure aspirator mixer, a control valve, either manually or motor operated, is located in the combustion air line (Figure 3.6). Opening or closing the valve increases or decreases the flow of air to the mixer. Usually the flow of air is not shut off completely but is turned down to a low-fire or minimum flow.

FIGURE 3.6: LOW-PRESSURE-ASPIRATOR PIPING

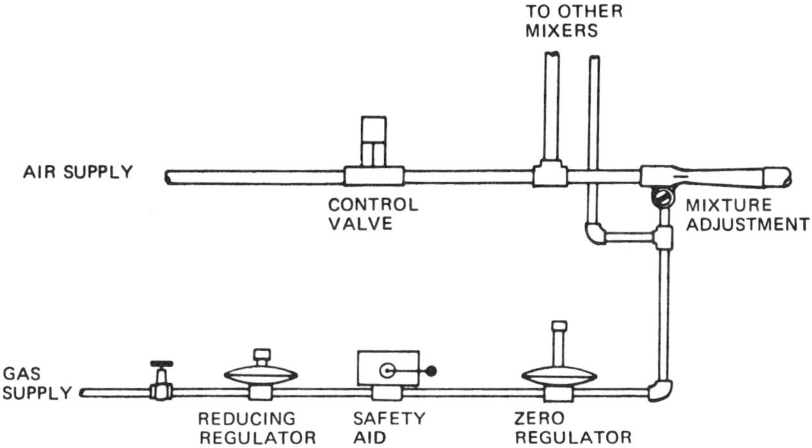

Source: TID-27600

High-pressure inspirators are controlled by a valve located in the gas line to the mixer (Figure 3.7). This valve can be either manually or automatically operated to control the temperature of the furnace. The air flow to the mixer will follow the gas flow to maintain a constant air/gas ratio in the mixture regardless of firing rate. Again high-pressure inspirators are usually controlled with a minimum flow setting on the valve.

FIGURE 3.7: HIGH-PRESSURE-INSPIRATOR PIPING

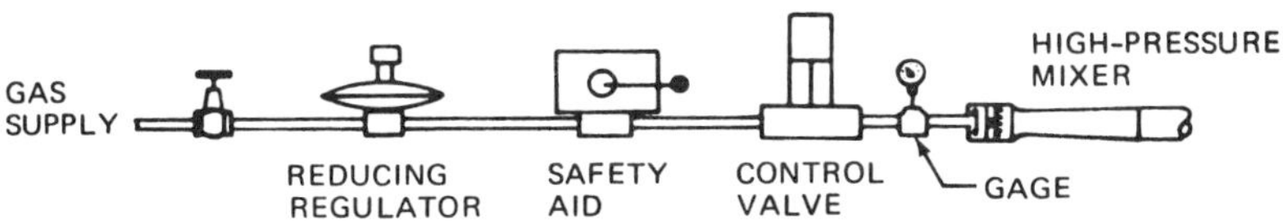

Source: TID-27600

In combustion systems incorporating mechanical mixers, the firing rate is usually controlled by a valve located in the air-gas mixture line (Figure 3.8). This valve can be either manually or automatically operated and is usually set for a low-fire or minimum flow condition. Since the flow rate is varied by the valve, the carburetor or blender on the mechanical mixer automatically controls the air/gas ratio to remain constant. The exact method of control of the air/gas ratio depends on the make of mixer used (zero regulator type shown in Figure 3.8).

FIGURE 3.8: MECHANICAL-MIXER PIPING

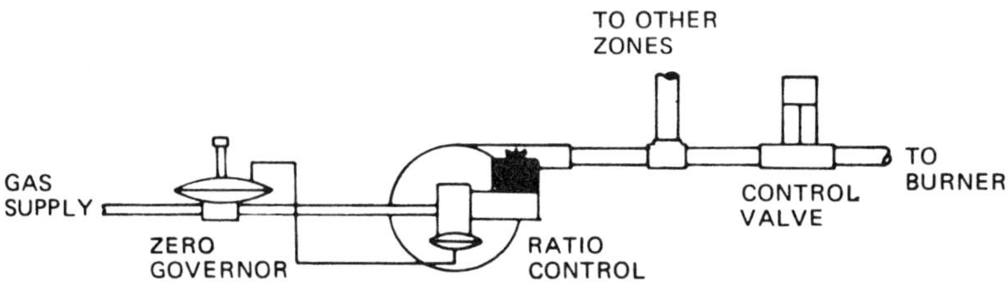

Source: TID-27600

Nozzle-Mix Burners

A nozzle-mix burner is a device in which the combustion air and fuel are introduced separately and mixed within the burner body just prior to ignition of the mixture. The nozzle-mix burner has separate air and fuel connections. A gas burner has only one air inlet, whereas an oil burner has an additional connection for the atomizing fluid (air and steam).

The nozzle-mix burner uses a torch, pilot, or spark plug for burner ignition. In small installations the burners are ignited manually with a propane torch or similar device. When a pilot is used, it uses a premix air-gas mixture; it may be "on" for the entire period of the burner operation (constant pilot) or it may be used in the beginning to light the burner and then turned "off" (interrupted pilot). The pilot can be ignited manually or by a spark plug in the pilot. Sometimes the burner is lighted directly by a spark plug, which is then turned off during the normal burner operation. Most burner manufacturers and insurance underwriters recommend lighting the burner at low firing condition for all types of ignition devices.

The nozzle-mix burners are designed to operate over a wide range of air/fuel ratios. In most applications the burners are operated at a fixed A/F ratio, and the burner control system is designed to maintain this ratio at all firing rates or fuel inputs. The exact A/F ratio depends on the type of application.

In many small furnaces and special applications, the furnace temperature must be varied over a very wide range, typically from 300° to 1800°F. The use of a special type of nozzle-mix burner, known as an excess-air burner, is common in such applications. In excess-air burners the combustion air is maintained constant at all times, and the fuel input is changed according to the heat and temperature requirements of the furnace. Obviously, this type of burner operation is highly inefficient and should be avoided except where process requirements cannot be met with constant A/F ratio type burners.

Typical Combustion Control Sytems: *Burner Control*—Most furnaces are divided into more than one zone. Each zone has more than one burner. The zone temperature is controlled by controlling the firing rate of the burners in each zone. The most common methods of controlling the zone temperature with constant A/F ratio type burners are on-off control, two-position or high-low firing control, and modulating or proportional control.

In an on-off control system, the burner is fired at the full or rated capacity when the zone temperature is below the set point, and it is turned off when the zone temperature is above the set point. For the constant-piloted burners, the pilots are kept "on" when the burner is turned "off" by the controller action.

In many furnaces the burners are operated at low-fire position by reducing the air and fuel supply to the burner instead of completely turning off the burner when the zone temperature exceeds the set point. These burners generally use a spark plug or an interrupted pilot, which is "on" only during burner lighting for the ignition. In such burners, when the zone temperature is below the set limit, the burners are operated at the full or rated firing capacity.

When a precise or closer temperature control is desired, a proportional controller is used to modulate the air and gas supply to the burner. Burner firing depends on the difference between the set point and the actual temperature.

The air and fuel piping and the control hardware depend on the type of control scheme used.

Air/Fuel Ratio Control—Three most commonly used methods for the A/F ratio control are balanced pressure, tandem-valve control, and volumetric flow control.

In the balanced-pressure control scheme, a control valve in the combustion air line is used to control the firing rate of the burner. In a typical burner system (Figure 3.9), a common pipe or header for air and one for fuel supply the air and fuel to one or more burners of a zone. The air flow is controlled by the action of the zone temperature controller. The fuel supply to the zone burners is controlled by a zero (or ratio) regulator in the fuel line. The zero (or ratio) regulator maintains the fuel pressure at the regulator equal to or proportional to the air pressure downstream of the control valve in the air line. An impulse or pressure load line is used to back-load the zero (or ratio) regulator. If the fuel (oil or gas) pressure is more than the air pressure, the impulse line transmits the direct air pressure to the regulator.

FIGURE 3.9: BALANCED-PRESSURE RATIO CONTROL SYSTEM FOR NOZZLE-MIX BURNERS

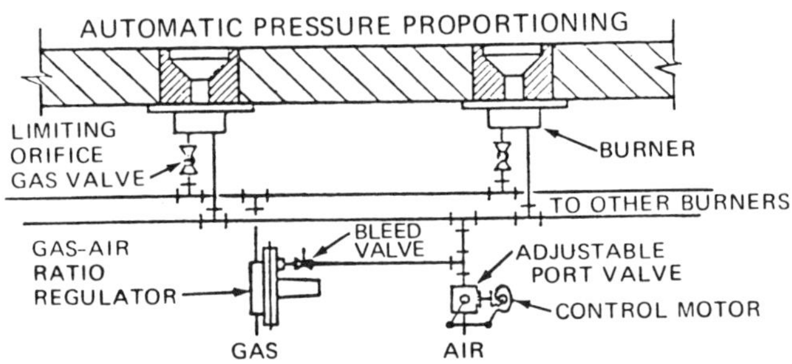

Source: TID-27600

However, if the fuel (normally gas) pressure is less than the air pressure, a bleeder valve in the impulse line reduces the regulator back-loading pressure. As the air pressure changes, the pressure signal, which is equal to or proportional to the air pressure, is transferred to the zero (or ratio) regulator, and it adjusts the position of a valve in the fuel line to make the fuel pressure proportional, or equal, to the air pressure.

For a precise control of the A/F ratio at all flows or firing rates of the burner, the resistance in the air and fuel line downstream of the control valve and the zero regulator, respectively, should remain constant. This means that the density, viscosity, and other properties of the air and of the fuel and the flow areas in the lines and the burner should not change when the burner firing rate is changed.

The tandem-valve control system (Figure 3.10) uses one control valve in the air line and one control valve in the fuel line. Both these valves are operated by a single control motor through proper linkages. The linkages and the valves are designed such that for any position of the motor shaft the air and fuel flows through the valves are proportional.

FIGURE 3.10: TANDEM VALVE RATIO CONTROL SYSTEM FOR NOZZLE-MIX BURNERS

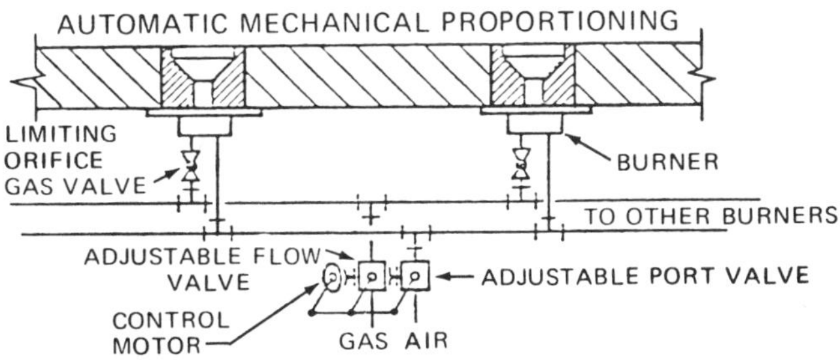

Source: TID-27600

In the volumetric flow control system, illustrated in Figure 3.11, accurately machined orifices are used in the air line and in the fuel line. The pressure drop through the orifices represents the air and fuel flow. The two sides of each of the orifices are connected to a ratio controller. When the zone fuel requirement changes, the controller operates a valve in the fuel line (or air line) to control the fuel or air input, respectively. This changes the pressure differential through the orifice in the corresponding line. The ratio controller senses this change in one of the lines and acts to reposition the air (or fuel) valve until the ratio of the pressure differential in the air line and in the fuel line is re-established to the preset or desired value.

For the most efficient burner operation, the A/F ratio to the burner should be maintained as close to the stoichiometric value as possible for a given application.

FIGURE 3.11: VOLUMETRIC-FLOW-CONTROL-TYPE RATIO CONTROL SYSTEM FOR NOZZLE-MIX BURNERS

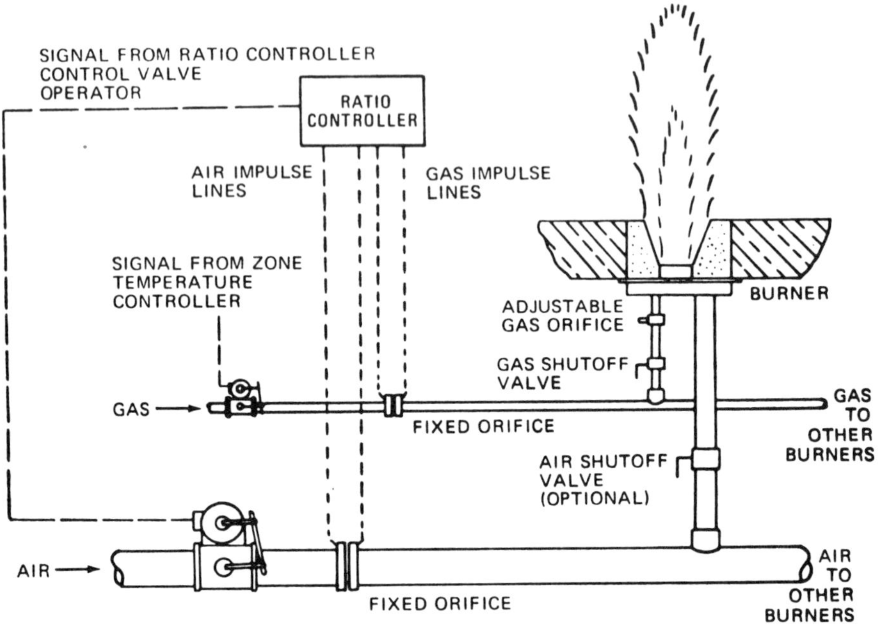

Source: TID-27600

Radiant-Tube Burners

Radiant tubes are used to heat the work indirectly so that there is no direct contact between the flue products and the work. The most commonly used shapes for radiant tubes are shown in Figure 3.12. The tubes can be straight, U-shaped, or W-shaped. In each case the burner is fired on one end of the tube, and the flue gases are discharged from the other end. In one particular type of tube, known as a trident tube, the burners are mounted on two outer legs, and the exhaust gases are collected in and discharged from a common middle leg.

Radiant-tube burners can be divided into two types: 1) push-through or pressure type and 2) pull-through or suction type.

In push-through type burners, the combustion air is supplied to the burners directly from the blower and at some pressure. Accordingly, the gases in the tube are slightly above atmospheric pressure. In pull-through type burners, the combustion air is obtained from the atmosphere by the suction at the burner end. The suction is created by an exhaust fan or a pressurized air jet located at the exhaust end of the tube.

FIGURE 3.12: MOST COMMONLY USED RADIANT-TUBE SHAPES

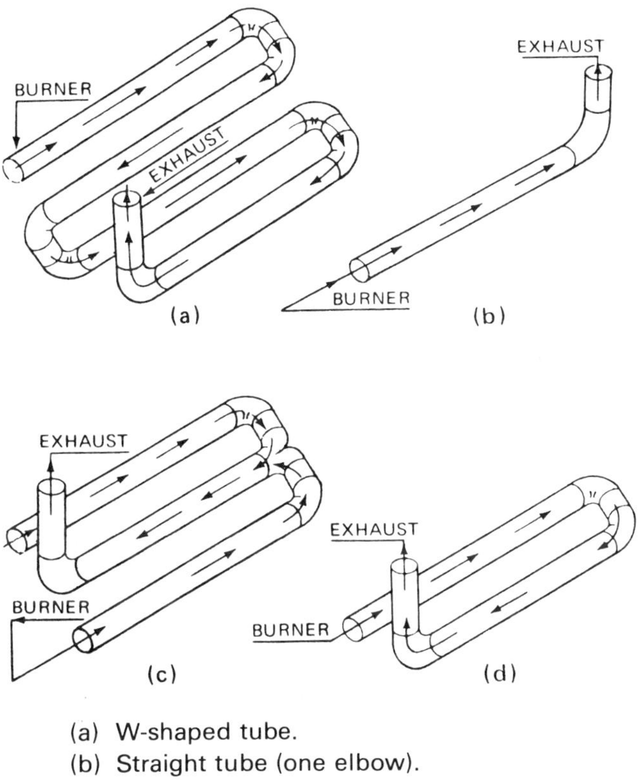

(a) W-shaped tube.
(b) Straight tube (one elbow).
(c) Trident tube.
(d) U-shaped tube.

Source: TID-27600

The burners can be gas fired or oil fired. In almost all applications a uniform tube temperature must be obtained over the entire tube length. Hence radiant-tube burners are designed to give a long, lazy, and luminous flame. Since the length of radiant tubes in different applications varies over a wide range and it is necessary to match the flame length with the tube length, many burner designs have some adjustment available to control the flame length. The nozzle-mix arrangement is almost exclusively used for radiant-tube burners.

A radiant-tube burner can be ignited manually with a gas torch or by a premix pilot or spark plug. The fuel input in the pilots is effectively used to heat the tube.

Typical Combustion Control Systems: Typically, a furnace using radiant tubes may be divided into more than one zone. Each zone may contain several tubes of the same size or of different sizes and shapes. The air and fuel supplies to the burners on these tubes are, usually, from the zone air and fuel supply lines, which include the control valves. All the tubes in a zone are fired at the same time, although not necessarily at the same heat input rate. The piping and location of control-sytem components depend on the type of burner and the mode of control system used.

Pull-Through (Suction) Burner—A typical pull-through burner system is shown in Figure 3.13. All presently available burners use a constant premix gas pilot with gas or oil firing. The fuel, oil or gas, is controlled by a valve in the zone supply line. In most cases there is no adjustment available for the individual burner in the fuel supply line. Although a shutoff valve between the zone control valve and the burner can be used for each burner, it should always be kept full open during normal operations. The burner input is preset by the supplier by proper sizing of the fuel nozzle.

The combustion air is controlled by regulating the suction at the burner. In most cases an eductor assembly is used to create the suction. The combustion air supply depends on the eductor air flow and pressure. The eductor spud or orifice is sized to give the correct A/F ratio at a given air pressure. When the fuel flow needs to be changed because of a change in fuel pressure or flow area, the eductor air-flow area or pressure must also be changed. In many designs there is some provision for adjusting the eductor air flow and hence the combustion air flow. A screw adjustment can be used to change the eductor air-flow area or an opening at the end of the tube can be controlled by a disk to introduce atmospheric air to the combustion products from the tube and to reduce the suction effect at the burner.

In some cases the suction at the burner end is created by connecting one or more exhaust ducts to a main exhauster that pulls the flue gases from the tubes. The suction to the tubes is controlled by adjusting a fresh-air damper connected to the exhaust duct.

FIGURE 3.13: TYPICAL PULL-THROUGH (SUCTION) BURNER SYSTEM

ON–OFF CONTROL SYSTEM
PULL-THROUGH RADIANT–TUBE BURNER

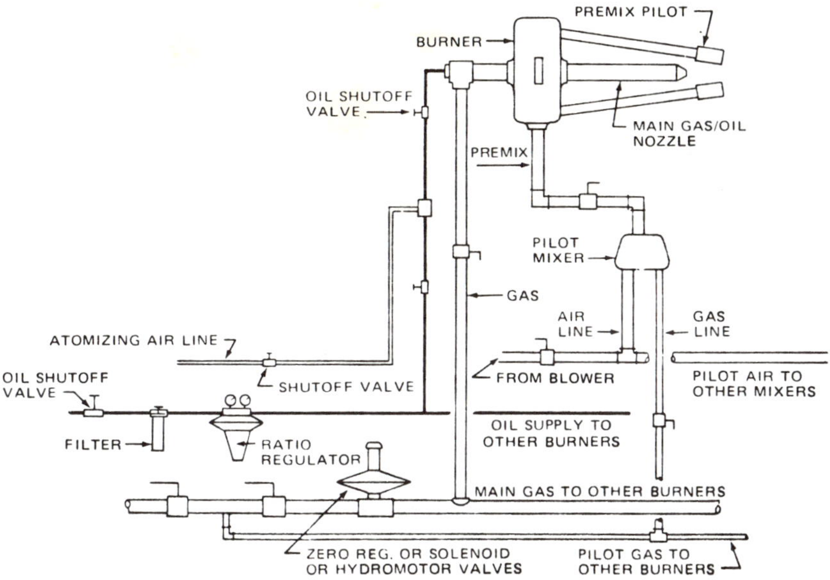

ON-OFF CONTROL SYSTEM
RADIANT-TUBE EDUCTOR ELBOW

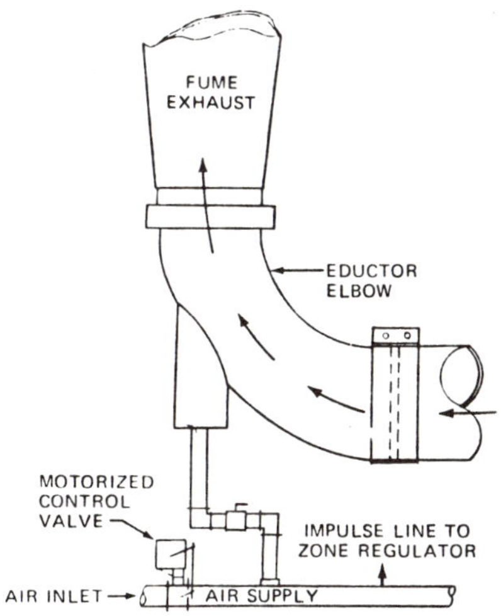

Source: TID-27600

In a typical system a control valve in the eductor air line for a zone is opened or closed by the controller action. For gas burners a regulator valve or some type of shutoff valve is used in the gas line to regulate the gas supply to the burner. An impulse or pressure load line is often used to back-load the regulator. When the eductor air-line valve is closed, the pressure downstream of the control valve reduces to almost zero, and this pressure signal closes the regulator valve in the gas line. This ensures that the eductor (and combustion) air and gas supplies to the burners are in harmony.

For oil burners the eductor air and oil supplies are controlled separately by installing a solenoid or a motor-operated valve in each line. Both valves are opened and closed simultaneously by the controller action. The atomizing air supply to the oil burner is either maintained constant or reduced by using a proper valve in the line.

For systems using an exhauster, the damper in the flue-gas passages and a solenoid or motor-operated shutoff valve in the fuel line are opened or closed simultaneously to control the burner input.

When the pull-through burner is used with the modulating control, the valve in the eductor air line is operated by a proportional controller, and the pressure signal, downstream of the valve, is used to control the zero regulator or ratio regulator in the fuel line. The system is similar to the one used with nozzle-mix burners.

Push-Through (Pressure) Burner—A push-through burner is basically a nozzle-mix burner in which the combustion air and fuel are supplied separately, but directly, to the burner. A typical radiant-tube system using push-through burners is shown in Figure 3.14. The fuel may be gas or oil. For oil firing an additional supply line is used to introduce the atomizing agent, which may be air or steam.

The burners can be operated with on-off control, high-low firing control, or modulating (proportional) control.

The A/F ratio control for the push-through burners is identical to the controls used for nozzle-mix burners. For push-through burners the balanced-pressure system is used almost universally.

For the most efficient burner operation, the A/F ratio to the burners should be maintained very close to the stoichiometric value, and the fuel combustion should be completed within the radiant tube. In most radiant-tube applications, the burner A/F ratio is regulated to obtain approximately 2% oxygen in flue gases from the tube.

FIGURE 3.14: TYPICAL PUSH-THROUGH (PRESSURE) BURNER SYSTEM

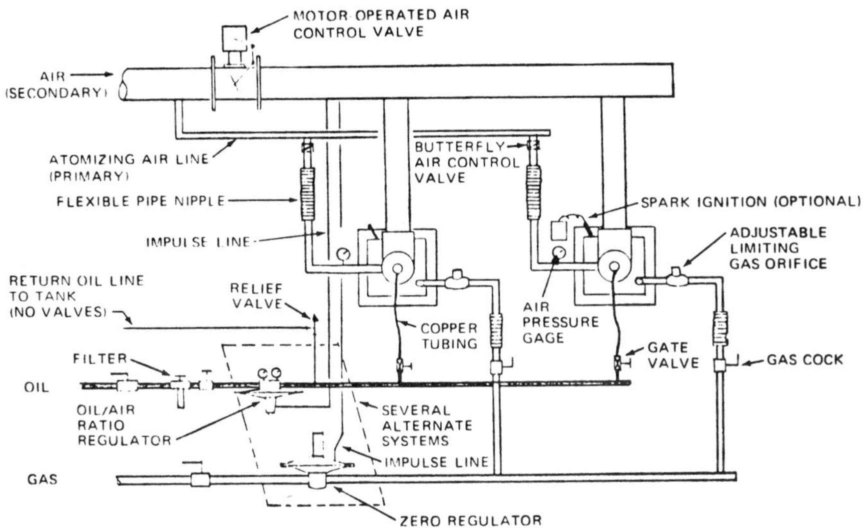

Source: TID-27600

OPTIMIZATION OF FUEL CONSUMPTION

Combustion and Air/Fuel Ratio

Unless favorable conditions are maintained in the region where combustion is taking place, the fuel may not release all its chemical energy and the energy released may not be used effectively in subsequent steps of the conversion path.

The amount of heat liberated in any combustion reaction depends on the relative amounts of fuel and air that are supplied and on the type of fuel. The amount of heat contained in a specific fuel is expressed in terms of Btu per cubic foot or per pound of the fuel. The relative amounts of air and fuel burned, represented by the air/fuel ratio (A/F), is expressed in cubic feet of air per cubic foot of fuel or pounds of air per pound of fuel. Heating values and A/F ratios for typical fuels are shown in Table 3.1.

For a given fuel the greatest liberation of heat will result when the amount of air supplied completely burns (or oxidizes) the fuel. This is referred to as perfect, or stoichiometric, combustion. The A/F ratio that results in perfect combustion is called the stoichiometric A/F ratio and is shown in Table 3.1.

TABLE 3.1: THEORETICAL AIR/FUEL RATIOS FOR COMMON FUELS

	Gross heating value, Btu/ft³ or lb	Theoretical stoichiometric air required		
		Standard ft³/ft³	Standard ft³/lb	Standard ft³/1000 Btu
Natural gas (Toledo)	1,050	10.16		9.67
Natural gas (Pueblo, Colo.)	980	9.19		9.38
Natural gas (Los Angeles)	1,084	10.26		9.46
Hydrogen	325	2.393		7.36*
Carbon monoxide	322	2.393		7.36*
Propane (gas)	2,520	23.925		9.49
Coke oven gas (Koppers)	590	5.28		8.94*
Producer gas (anthracite coal)	150	1.12		7.46*
Butane	3,200	31.09		9.71
Blast furnace gas	92	0.68		7.39*
No. 2 fuel oil	19,500		192.3	9.80
No. 6 fuel oil	18,300		181.82	9.93
Low-ash bituminous coal	14,490		143.64	9.91
High-ash bituminous coal	12,780		127.0	9.93
Lignite dry	10,082		102.9	10.20
Lignite wet	7,063		72.6	10.20

*Note the variation from the simple rule of thumb that gives 10 ft³ of air per 1000 Btu heat release.

Source: TID-27600

If an insufficient amount of fuel is supplied for the combustion reaction, combustion is said to be "lean." Therefore, when an excess of air has been supplied, excess oxygen can be detected in the flue gases. The heat released in this reaction is less than that in stoichiometric combustion since some heat must be used to heat the extra nitrogen and oxygen that did not participate in the reaction.

When more fuel is supplied than is required for stoichiometric combustion, the reaction is referred to as "rich," or "air deficient," combustion. Both H_2 and CO therefore will appear in the flue gases. This combustion is inefficient since CO and H_2 are combustibles and would liberate additional heat going to carbon dioxide and water vapor if more oxygen were available. Thus the heat released in rich combustion is less than that in stoichiometric combustion because not all the fuel has been used. In addition, because of its high toxicity, CO is dangerous, and care must be exercised when working with equipment running at fuel-rich conditions.

Excess air is required in practical combustion equipment primarily because of the difficulty of achieving satisfactory mixing between the fuel and the air. In the usual process of combustion, oxygen is not supplied directly but rather as a constituent (21% by volume) of atmospheric air. Thus some 79% of the material that enters the furnace with fuel consists of nitrogen or other inert gases which pass through the furnace (and through the process of combustion) without making any thermal contribution but absorbing, in sensible heating, a significant amount of the energy liberated from the fuel. The excess air acts as a diluent and thereby reduces the temperature of the combustion process; in many cases this reduction is necessary and desirable for structural and metallurgical reasons, but thermodynamically it is always disadvantageous.

Available Heat Effects

Only part of the heat liberated in combustion can be used for heating a load because the flue gases will leave the furnace at elevated temperatures. The percentage of the heat of combustion that can be used is called "available heat."

Factors affecting the available heat of a given process are final flue-gas temperature, furnace and burner design, and ratio of air to fuel.

The final flue-gas temperature is, of couse, defined by the particular heating process and the required work temperature, and these conditions cannot readily be changed to improve fuel efficiency. The use of heat-recuperation equipment is beyond the scope of this book since it is not directly related to burner adjustments and combustion control.

Although improvements in available heat can be effected by improvements in furnace and burner design, improved designs would require new process equipment and thus cannot be considered as short-term or inexpensive solutions.

If combustion is not stoichiometric or near stoichiometric, however, significant improvement in the amount of heat available to the work can be made by relatively simple and inexpensive adjustments in the

A/F ratio to achieve the stoichiometric ratio. The dominant factor governing the amount of heat that is transferred to the work is the temperature difference between the work and the combustion products. The greater the temperature difference, the greater the amount of heat transferred.

Since heat release, and therefore flue-gas temperature, are highest at stoichiometric combustion, more heat will be available for transfer as stoichiometric combustion is approached. This is demonstrated in Figure 3.15, where the theoretical flame temperature of natural gas and air is plotted as a function of the A/F ratio. As the A/F ratio approaches the stoichiometric ratio, the flame temperature increases. It decreases at ratios above stoichiometric. The highest flue-product temperatures are therefore obtained at or near the stoichiometric A/F ratio.

FIGURE 3.15: THEORETICAL FLAME TEMPERATURE vs AIR/FUEL RATIO FOR NATURAL GAS

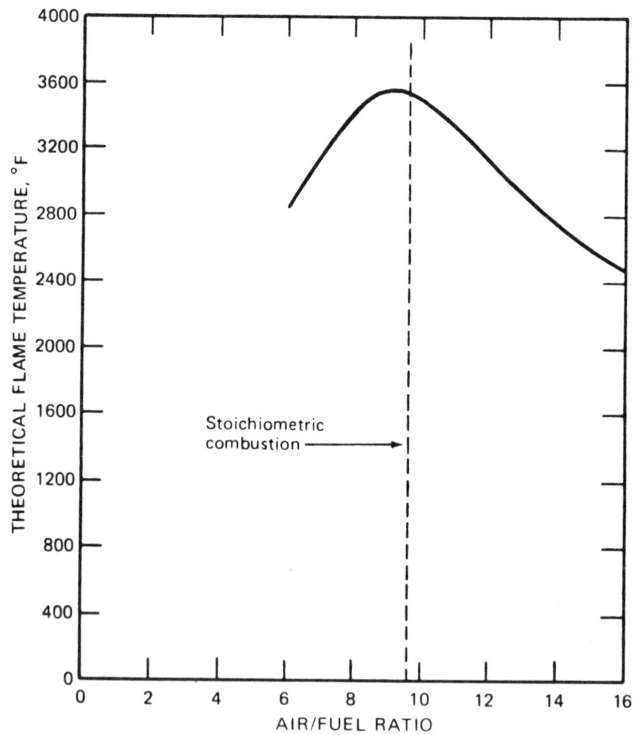

Source: TID-27600

Effect of Combustion on Fuel Use

In Figure 3.16 the available heat of natural gas is plotted as a function of flue-gas temperature for various amounts of excess air. Stoichiometric combustion is represented by the 0% excess air curve. This figure demonstrates the effect nonstoichiometric combustion has on fuel use.

FIGURE 3.16: PERCENT AVAILABLE HEAT OF NATURAL-GAS vs FLUE-GAS TEMPERATURE WITH VARYING EXCESS AIR

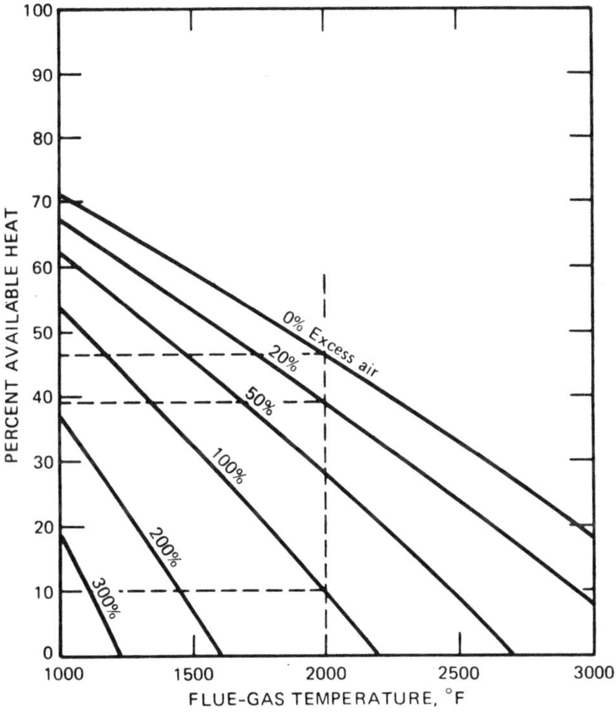

Source: TID-27600

If a particular heating process requires the work to be heated to 2000°F, Figure 3.16 shows that when operating at stoichiometric conditions (0% excess air) the available heat is 47% but that when operating at 20% excess air the available heat is only 39%. Therefore, to obtain the same amount of heat at 20% excess air as is obtained at stoichiometric combustion, one must increase the amount of fuel used by 0.47/0.39, or 1.20; i.e., 20% more fuel is required when operating at 20% excess air. When operating at 100% excess air, the available heat is reduced to 10%. This represents a fuel increase of 0.47/0.10, or 4.7 times that required at stoichiometric conditions.

Similar increases in gas consumption result at substoichiometric or fuel-rich conditions. This is reflected in Figure 3.17, where the percent fuel savings is plotted as a function of both the percent combustibles and the percent oxygen in flue product at various temperatures. Fuel use is therefore a minimum at stoichiometric ratios.

FIGURE 3.17: FUEL SAVINGS vs PERCENT OXYGEN AND COMBUSTIBLES IN THE FLUE PRODUCTS

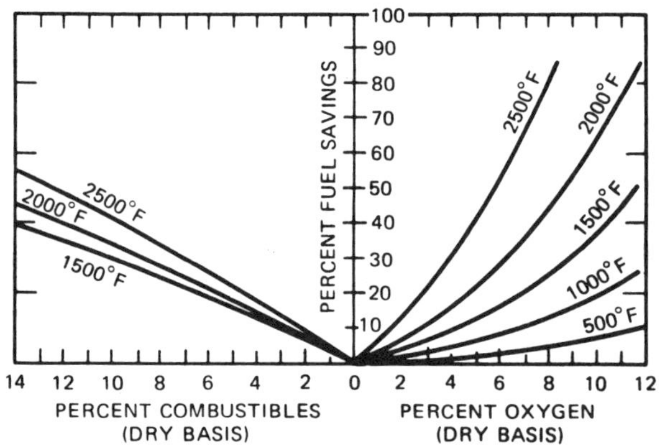

Source: TID-27600

The preceding relationships can be expressed in a slightly different way by considering the combustion equation for CH_4:

$$CH_4 + 2O_2 + 7.52N_2 \rightarrow CO_2 + 2H_2O + 7.52N_2 + \text{Heat}$$

If zero excess air is present, the volumes of the combustion products become

	Percent
Carbon dioxide (CO_2)	7.00
Water (H_2O)	14.01
Nitrogen (N_2)	78.99
Oxygen (O_2)	0
Total volume	100.00

If 10% excess air is introduced into the combustion process, the volumes of the combustion products now become

	Percent
Carbon dioxide (CO_2)	6.37
Water (H_2O)	12.73
Nitrogen (N_2)	78.56
Oxygen (O_2)	2.34
Total volume	100.00

When 10% excess air is introduced into the combustion process, 2.34% oxygen by volume is lost. Assuming that the same amount of heat (2000°F) is desired as was used previously (Figure 3.16) and that the available heat at zero excess air is 47%, and at 10% excess air, approximately 45%, then the amount of fuel must be increased by 0.47/0.45, or 1.023. That is a 2.3% increase in fuel; i.e., CH_4 is required to consume the 2.3% oxygen present in the combustion products. These calculations also could be extended to calculate the amount of CO_2 that would be present in the combustion products if insufficient oxygen were present to completely consume the methane. Thus analyzing the products of combustion is a good way to control the combustion process.

Process Requirements

Although the benefits of stoichiometric combustion are obvious, some attention must also be given to process requirements. There are instances when a process may require combustion to be at non-stoichiometric conditions. In heating steel, for example, the presence or absence of oxygen in the flue products could be a process requirement.

It may be necessary, for example, to burn at substoichiometric combustion to reduce the amount of oxidation, or "scale," on the work surface. On the other hand, surface oxidation may be highly desirable since a later manufacturing process, such as pickling, may be shortened by the existence of scale.

In this case burning with excess air would be required. Excess-air combustion could also be necessary to reduce flame temperatures to avoid local overheating or melting of the work.

Boiler Testing

The material in this chapter was excerpted from a report by KVB, Inc. (PB-264 543).

To reduce NO_x emissions and improve boiler operating efficiencies, it is necessary to determine how NO_x and efficiency vary as boiler adjustments are made. To successfully accomplish this requires a preliminary boiler inspection to assess the existing condition of the furnace, burner equipment and firing parameters. Completing any necessary repairs and adjustments will insure that the subsequent efficiency and emission results are representative of the most favorable, as-designed, condition of the boiler. In some cases, this first step will itself directly lead to efficiency improvements.

The other essential part of efficiency improvement/NO_x reduction involves the actual testing to establish the optimum boiler adjustments. These include boiler instrumentation, visual furnace observation, and stack emission measurements. Some of these areas involve procedures and equipment which have not traditionally been a part of routine boiler operating practice.

PRELIMINARY BOILER INSPECTION

Even when boiler equipment is in poor operating condition, "best" boiler settings for improved efficiency and reduced NO_x emissions can still be found under these conditions using the standard test procedures. However, it is stressed that these efficiency improvements obtained under a deteriorated state of the boiler can be substantially less than the improvements achieved when the boiler equipment is in proper working order. To attain maximum fuel savings and lowest stack emissions, it is essential that the condition of the boiler be examined prior to testing and that any necessary repairs or maintenance be completed.

There are many maintenance items which can affect boiler efficiency and they are described later. Both the manufacturer's boiler operating and maintenance manual and a list of maintenance items should be used to ready the boiler for the efficiency and emission tests. Prior to making extensive adjustments and/or repairs, the manufacturer's field service department or other combustion consultants should be consulted. Some of the more common items to be included in the preliminary boiler inspection are summarized here.

Burners

For oil firing, one should make sure that the atomizer is of proper design and size for the type of oil and burner geometry. (There may have been a mix-up between boilers when oil tips were removed at the cleaning bench.) The oil tip passages and orifices should be inspected for excessive erosion and any coke or gum deposits removed to assure a proper oil spray pattern. Oil temperatures at the burner should be checked to see that they are at recommended levels. Atomizing steam, if used, must be at the proper pressure. Burner diffusers should be inspected to make sure they are not burned off or broken and are properly located with respect to the oil gun tip. Also the oil gun should be checked so it is positioned properly within the burner throat. Any damaged or missing burner throat refractory should be replaced. Oil strainers should be in place and clean.

When firing natural gas, the gas injection orifices should be inspected and it should be verified that all passages are unobstructed. Filters and moisture traps should be in place, clean, and operating to prevent plugging of gas orifices. Proper location and orientation of diffusers, spuds, gas canes, etc. should also be confirmed. Any burned off or missing burner parts should be looked for.

Pulverized coal burner components such as pulverizers, feeders, conveyors, and primary/tempering air ducts must all be working properly.

Coal fineness should be within recommended limits. All coal pipes should be cleared of any coal and coke deposits. Burner parts should be checked for any signs of excessive erosion or burn-off. For stoker boilers, the grates must not be excessively worn causing poor burn-out in the bed or high carbon carryover. Spreaders should be checked to ensure they are working properly and are in the correct location. The proper positioning of all air proportioning dampers should be confirmed. Proper coal sizing is also important. To minimize unburned carbon losses, the cinder reinjection system must also be operating correctly.

Combustion Controls

All fuel valves should be inspected to verify proper movement and clean internal surfaces. Gas and oil valve surfaces may erode with time. There should not be excessive "play" in control linkages or air dampers. Fuel supply inlet pressures to all pressure regulators should be adequate to assure constant regulator outlet pressures for all firing rates. Atomizing steam or air systems must also deliver proper flows. Control elements which fail to respond smoothly to varying steam demand should be corrected. Unnecessary cycling of firing rate due to improperly adjusted regulators or automatic master controllers can waste fuel. All gauges should be functioning and calibrated to aid in identifying any control problems as they occur. All safety interlocks and boiler trip circuits must be operable.

Furnace

Inspect gas-side boiler tube surfaces for excessive deposits and fouling. These lead to higher stack temperatures and lower boiler efficiencies. Poor firing conditions may be the cause of tube deposit problems but proper operation of soot-blowing equipment should also be checked. Periodic cleaning of tube surfaces may be a practical solution when burners and soot blowers are found to be operating as designed.

Any leaks in the boiler gas passages and baffling should be repaired. Furnace refractory and insulation should also be inspected and any casing leaks and cracked or missing refractory repaired.

Furnace inspection ports must be cleaned and operable since visual flame observations will be an essential part of the boiler testing. Inspection ports should provide a view of the burner throat, furnace walls and leading convection passes.

STACK INSTRUMENTS

Boiler adjustments to improve efficiency and reduce nitrogen oxide emissions cannot be attempted without proper stack instrumentation.

To successfully carry out these tests in a safe manner, it is necessary to measure the following quantities at the stack: excess oxygen (or carbon dioxide), carbon monoxide, oxides of nitrogen, opacity (smoke density), and stack temperature.

These measurements must also be accompanied by visual furnace observations to assure acceptable flame conditions.

Excess Oxygen

Excess oxygen concentration in industrial boiler stacks can vary from a fraction of 1 to 10% or more depending on boiler design, type of fuel, burner adjustments and firing rate. The lowest practical excess O_2 is usually the preferred condition for highest efficiency; NO_X emissions also are generally reduced by operating with lower excess O_2.

Oxygen can be measured using portable analyzers when the boiler is not already equipped with O_2 meters. Reasonable accuracies can be obtained with an Orsat analyzer by an experienced technician using fresh chemicals. Other handheld chemical-absorbing type analyzers and "length of stain" (color-sensitive tubes) devices are commercially available. Portable or mounted electronic instruments with high accuracies are also available.

Whether using a built-in O_2 analyzer or a portable instrument, its proper calibration is essential for accurate readings and successful boiler adjustments. The recommended calibration procedure for the instrument should be followed and the frequency of calibration adjusted as necessary to minimize calibration "drift." One should start with a fresh calibration when beginning the boiler tests. When calibration gases are required, newer gases should be cross checked when delivered, with older ones to catch any errors in the calibration gas itself. For analyzers involving absorbing chemicals (such as Orsat analyzers), the absorbing chemicals should be fresh and a routine established for renewing chemicals based on the number of samples analyzed and the age of the chemicals. These are general remarks concerning instrument calibration and they also apply to the other stack measurements.

Carbon Monoxide

On natural gas fired boilers, CO is the primary indicator of incomplete combustion and will usually determine the lowest practical excess O_2 for the boiler. Carbon monoxide concentration in the stack gases should not exceed 400 ppm once the final boiler adjustments are made. Some city ordinances, industry codes and insurance organizations require that CO not exceed this value. For purposes of testing, occasional CO levels of up to 1,000 or 2,000 ppm can be acceptable provided adequate boiler monitoring and flame observation are possible to assure stable conditions. Caution should be used at these conditions. Further reducing the excess air will result in very rapid increases in CO and other combustibles which can lead to smoking, flame instability, furnace pulsation and boiler explosions.

Carbon monoxide measurements on oil and coal fired boilers are not generally mandatory since smoking or excessive carbon carryover will usually precede any substantial CO emissions. However, this cannot always be anticipated. For example, high CO levels have been measured in cases where burner equipment has deteriorated or malfunctioned (burned off impellers, plugged oil tips, insufficient overfire air, etc.). High CO emissions can also be encountered at high excess O_2 levels where the air flow has been increased to the point of influencing flame stability. For this reason, it is recommended that CO measurements be included on all boilers to assure that there are no burner equipment problems complicating the efficiency and NO emission test results.

The CO analyzer should be capable of measuring up to 2,000 ppm and should also have sufficient sensitivity to measure down to the 100 ppm range. Orsat analyzers have traditionally been used for CO determinations but difficulties in accurately reading the lower CO concentrations (less than 1,000 ppm) has spurred the use of handheld "length of stain" type detectors which reportedly provide the necessary sensitivities and measurement range. Portable or mounted electronic instruments are also available.

Oxides of Nitrogen

To assist in selecting NO_X instrumentation with the proper range of measurement, Table 4.1 shows the range of NO_X emissions found during field testing at industrial boilers throughout the country (1)(2)(3). These NO_X emission levels generally correspond to boiler firing rates near 80% of rated load. Since NO_X usually increases with firing rate, these values are close to the highest NO_X emissions found during normal operation. Note that higher NO_X emissions also generally occur on boilers equipped with combustion air preheaters.

TABLE 4.1: TYPICAL RANGE OF NO_X EMISSIONS

Fuel Type	Without Air Preheaters	With Air Preheater
	ppm	
Coal		
Pulverized	300-600	400-800
Stoker	200-400	250-450
Cyclone	800-1100	900-1500
Fuel Oils		
#2	50-250	100-300
#5 (PS 300)	200-400	200-600
#6	200-400	200-600
Natural Gas	50-200	100-400

Source: PB-264 543

As is discussed in another chapter, total NO_X emissions from industrial boilers consist mainly of nitric oxide (NO) with the remainder made up of small quantities of nitrogen dioxide (NO_2). Since NO comprises more than 90% of total NO_X and accurate NO_2 measurements are difficult, NO measurements are often taken as representative of total NO_X emissions from the boiler. There are a number of commercially available electronic instruments capable of measuring NO in concentrations typical of industrial boilers. They may differ significantly in measurement method, accuracy and price. Selection should be made with the guidance of a reliable vendor. Less expensive hand-held "length of stain" detectors sensitive to total NO_X (NO plus NO_2) are also available.

To properly interpret NO_X emission results, the measured NO_X reading obtained from the analyzer must be corrected to account for diluting effects of excess air in the flue gas. This is important when boiler excess air is varied and it is desired to know the resulting effect on NO_X emissions. For example, as excess air is raised, NO_X formation in the furnace may increase, but at the same time the additional air will dilute the flue gas and measured NO_X readings at the stack may actually be reduced.

To illustrate this point further, suppose a large air fan is installed at the boiler outlet to inject large quantities of outside air into the flue gas. The operator could then adjust the fan to dilute the stack gases and obtain any desired NO_X concentration less than the furnace outlet concentration.

However, if NO_X readings are corrected to some standard excess air (or excess O_2) condition, the "corrected" NO_X values will be the same regardless of how much diluting air is added. The purpose of correcting NO_X readings is not to prevent the use of diluting air fans or other similar devices. The primary purpose is to provide a consistent basis of comparing NO_X emissions and to eliminate the diluting effect of excess air when making these comparisons.

The accepted industry practice is to correct NO_X readings to a standard condition of 3% excess O_2. To do this, the measured NO_X reading in ppm (measured on a dry or moisture-free basis) is multiplied by an O_2 correction factor which is easily calculated. To determine the O_2 correction factor, divide 18 by the quantity 21 minus the percent excess O_2 measured in the stack, as shown below:

$$O_2 \text{ correction factor} = \frac{18}{21 - \%O_2}$$

For example, if the NO_X reading in the stack is 200 ppm and the excess O_2 at the same time is measured as 5.0%:

Corrected NO_X equals raw NO_X multipled by the O_2 correction factor, or,

$$\text{Corrected } NO_X = (\text{raw } NO_X) \times (O_2 \text{ correction factor})$$

$$= 200 \times \frac{18}{21 - 5.0}$$

$$= 200 \times \frac{18}{16}$$

$$= 200 \times 1.125 = 225 \text{ ppm}$$

The corrected NO_X will not necessarily remain constant as boiler excess air is varied (since NO_X formation is not constant). Raw NO_X readings must be obtained at each O_2 level and then corrected for the corresponding O_2.

Stack Opacity (Smoke Density)

Smoking with oil and coal fuels is a certain indication of flue gas combustibles or unacceptable flame conditions and should always be avoided. Some oil and coal fired boilers (especially larger capacity boilers) are equipped with a smoke detector which can be very useful in providing indications of poor stack conditions when properly calibrated. Ultimately, acceptable stack conditions are always confirmed by visual observation of the stack.

Accurate smoke measurements can be made on smaller boilers using inexpensive portable hand pump, filter paper testers. These provide spot check measurements but cannot be used for continuous stack monitoring. These devices use the smoke spot number or ASTM smoke scale (ASTM D2156) and can be very helpful in setting up optimum boiler operating conditions.

Stack Temperature

Tube deposits and fouling on the external tube surfaces of a watertube boiler or similar gas side conditions in the gas tubes of a firetube boiler will inhibit the absorption of heat in the boiler and lead to lower efficiencies. This condition will be reflected in high flue gas temperatures when compared to "clean" conditions at a similar firing rate and boiler excess air. The resulting loss in boiler efficiency can be closely estimated on the basis that a 1% efficiency loss occurs with every 40°F increase in stack temperature. Water side deposits resulting from inadequate water treatment would also eventually lead to higher stack temperatures; however, tube failures due to overheating generally occur before any substantial efficiency losses are evident.

Stack temperature measurements are an easy and effective means for monitoring boiler tube cleanliness conditions. Stack temperatures should be periodically compared to values obtained during start-up or following a boiler tube wash to determine any deviations from "clean" baseline temperatures. Since stack temperature usually increases with firing rate and excess air, these comparisons should be performed at similar boiling operating conditions.

In the absence of any previous data, the stack gas temperatures will normally be about 150° to 200°F above the steam temperature in a saturated steam boiler at high firing rates. This does not apply to boilers equipped with economizers or air preheaters. The boiler manufacturer should be able to supply a normal range of stack temperatures for the particular boiler design. When higher stack temperatures are measured, boiler tube cleaning may be warranted.

FLAME APPEARANCE

The appearance of an industrial boiler's flame can provide a good preliminary indication of combustion conditions. It is difficult to generalize the characteristics of a "good" flame since there is a certain amount of operator preference and variations due to burner design involved. This is especially true for stoker-fired coal boilers. For other types of combustion equipment, flames of a definite appearance have usually been sought. Short, bright, crisp, and highly turbulent oil and pulverized coal flames have been desired. Blue, slightly streaked, or nearly invisible flames have been sought for gas fuels. (However, operation with low NO_X emissions at reduced excess O_2 levels may result in a different flame appearance.) Stability of the flame at the burner and minimum furnace vibration are also universally desired. For underfed stoker boilers, an even bed and absence of carbon streamers are important criteria.

All too often, good flame appearance is achieved by operating with excess O_2 levels higher than necessary for safe, clean firing. Reducing the excess O_2 to the lowest practical levels for the boiler is the primary concern here.

At the other extreme, firing with insufficient excess O_2 has also been encountered at industrial boiler installations. This condition is usually limited to natural gas firing where very high combustibles can occur with improper burner settings and inadequate combustion quality checks. This description is intended to assist in correcting both conditions.

When firing with the lowest practical excess O_2, approximately the same amount of heat energy is released in the furnace for a given amount of fuel heat energy input. However, this process may take a longer period of time and utilize more furnace volume before the fuel is completely burned. The result of low excess O_2 firing is a flame which may have the following typical characteristics:

> Flames that actually grow in volume and more completely "fill" the furnace.
>
> Flames exhibiting a lazy "rolling" appearance. Instead of intense, highly turbulent flames, low O_2 flames may appear to flow somewhat slower through the furnace.
>
> The overall color of the flame may change as ex-

cess O_2 is lowered. Natural gas flames become more visible or luminous with yellow or even slightly hazy portions. Coal and oil fires become darker yellow or orange and may appear hazy in parts.

These characteristics are, for the most part, contrary to flame conditions traditionally desired by industrial boiler operators for clean, dependable firing. While this might seem discouraging, it is stressed that safety, reliability, and low particulate and soot emissions can still be achieved with low O_2 firing. This is true only if the required stack measurements are made and recommendations in these guidelines are followed. It should be mentioned that in many cases, firing with low excess O_2 will not necessarily produce any drastic changes in flame appearance.

If the boiler is already being fired with the lowest practical excess O_2, these guidelines will still be of benefit by helping to assure that high efficiencies and low NO_x emissions will be maintained.

STACK SAMPLING

To obtain measurements of excess O_2, CO, NO_x, and smoke, most of the analyzers previously mentioned require that a sample of flue gas be withdrawn from the stack and delivered to the analyzer. Some analyzers require a continuous flow of sample until a "steady" reading is obtained, while others require only a small volume of sample at one time. An exception is analyzers mounted "in-stack" with detector elements located in the flue gas stream. Regardless of the analysis technique, it is essential that the portion of the stack gases analyzed be a representative sample of the bulk of the stack gas flow.

The location of the sampling site at which the sample of flue gas is obtained is as important as the selection of proper measurement devices. Air leakage into the gas ducts on negative draft boilers can dilute the flue gas, and consequently, measurements will not give true indications of furnace conditions. Air leakage in air preheaters poses the same problem. When selecting a sampling site, it should be chosen upstream of the air preheater or any known air leaks when possible. All known air leaks upstream of the sample location should be sealed.

The sample site should also be removed from an area where the flue gas is highly turbulent, such as immediately downstream of bends, dampers, or induced draft fans. Flue gases can stratify or form "pockets" which can lead to errors especially when samples are withdrawn from a single point in the duct. When a single-point sample probe is used, it is recommended that readings at several points in the duct be compared to determine the most representative probe location. When existing sample ports are not adequate, it is well worthwhile to drill or cut out new ports to obtain accurate, reliable measurements.

Flue gas temperatures are also subject to stratification in the ducts and a representative location of the thermometer or other temperature sensors should be verified. The location should be close to the boiler outlet since temperature losses can occur in the flue gas ducting, especially in uninsulated sections.

BOILER TEST PROCEDURES

The two major aspects of the boiler testing are 1) determining the optimum boiler firing conditions over the turndown range, and 2) making the burner adjustments necessary to maintain these optimum conditions during normal automatic operation.

The second part of the testing is very dependent on the particular design of the boiler controls. To successfully accomplish the burner adjustments will require a working knowledge of the control system. For example, its design will dictate to a large extent the number of firing rate conditions necessary to be examined. An adjustment at one firing rate may affect conditions at other firing rates and these effects must be recognized and anticipated. It would not be desirable to improve efficiency and NO_X emissions at one firing rate only to create poor conditions at another.

Appropriate methods for varying burner excess air are described in another chapter. Again, they are dependent on the design of the burner and controls and can involve any number of air flow control devices such as forced draft fan inlet dampers, windbox dampers or stack draft dampers. Final adjustments to achieve the correct excess O_2 can involve adjustments to jackshaft control linkages, fuel valve cam profiles, fuel/air set points, etc.

The key word in applying combustion modifications is "Caution." One should be careful while making any adjustments and know at all times the impact on fuel flow, air flow, or the control system. All boiler safety interlocks and trip circuits must be functioning. Plant engineering personnel or the boiler manufacturer should be consulted if there is any uncertainty about the procedures or expected outcome of any adjustment. Flame appearance can give many clues as to combustion conditions—fires should be observed, but flame appearance should not be the only clue.

As described previously, flame appearance can change for low NO_X operation. Boiler instrumentation and the stack should be watched carefully while making changes. If in doubt, one should always check for combustibles (CO) in the flue gas. Conducting these tests may require additional manpower so that controls and instrumentation, flame appearance, and stack conditions can be monitored simultaneously during adjustments. All personnel should be familiar with the test objectives and fully instructed regarding their part in the test.

DATA RECORDING

To obtain maximum benefit from the boiler tests, all pertinent test data should be recorded. The permanent record of boiler operating conditions and stack measurements will not only document the boiler's efficiency and emission characteristics, but also enable future comparisons to help diagnose any efficiency or emission problems. The test data should be recorded on prepared data sheet forms and include the following items:

1) Identification of boiler, fuel type, date of tests and names of operating personnel involved.
2) Steam, feedwater and fuel conditions (flow rates, pressures and temperature) to document boiler firing rate and steam generation.
3) Combustion control position and burner settings.
4) Furnace pressures, temperatures, and damper settings.
5) Stack measurements (O_2, CO_2, CO, NO_X, smoke, temperature). Make note of sample probe position.
6) All relevant comments on flame appearance, carryover and furnace conditions.
7) Record any new permanent changes made to combustion controls or burner settings.

The actual boiler readings included will of course depend on the available instrumentation. Adequate data must be obtained so that exact boiler operating conditions can be repeated for future comparative purposes. A sample data sheet is shown in Figure 4.1, but additions or deletions will most likely be necessary for each particular boiler. It may be worthwhile to review the final form of the data sheet with the engineering staff and incorporate any additional data entries that may be of mutual interest.

FIGURE 4.1: SAMPLE DATA SHEET

NO_x REDUCTION TEST

PLANT ______________ TESTS BY ______________

BOILER NO. ______________ FUEL ______________ DATE ______________

Test No.					
Time					
Steam Flow, lb/hr					
Steam Pressure, psi					
Steam Temp., °F					
Fuel Flow, cfm, gpm, lb/hr					
Fuel Pressure, psi					
Fuel Temp., °F					
Atomizing Pressure					
Combustion Air Temp., °F					
Flue Gas Temp., °F					
Windbox Pressure, "H_2O					
Furnace Pressure, "H_2O					
Stack Pressure, "H_2O					
Fan Settings					
Air Register Settings					
Burner Positions					
Smoke Density					
O_2, %					
CO, ppm					
NO_x, ppm					
Flame Appearance: NOTES:					

Source: PB-264 543

Readings should be recorded only after steady boiler conditions are reached. This is usually indicated by steady stack temperature, fuel input, steam conditions (pressure and temperature and drum level). Steady excess O_2 readings in the stack are a good indication that fuel and air flows have stabilized.

It is very desirable that these tests be conducted at normal steam pressures. This will assure that stack temperatures and furnace temperatures are representative of normal operating conditions. Since it will usually be necessary to control the boiler firing rate manually during the tests to obtain stable conditions, this may pose some problems in satisfying normal steam demands. When alternate steam generating capacity is available, the loading system at other boilers should be modulated to maintain constant pressures. When this is not possible, it may be necessary to make provisions to dump unneeded steam or temporarily interrupt plant processes.

REFERENCES

(1) *Field Testing: Application of Combustion Modifications to Control Pollutant Emissions from Industrial Boilers—Phase II,* Report No. EPA-600/2-76-086a, April 1976 (PB-253 500).

(2) *Assessment of the Potential for Energy Conservation Through Improved Industrial Boiler Efficiency,* Final Report—Volume I, FEA Contract No. C-04-50085-00, October 1976 (PB-262 576).

(3) *Systems Evaluation of the Use of Low-Sulfur Western Coal in Existing Small and Intermediate Sized Boilers,* Interim Report No. 1, by KVB, Inc., Report No. 8800-140 under Contract No. EPA 68-02-1863, June 1975.

Reduction of NO_x and Other Impurities

The material in this chapter was excerpted from reports by KVB, Inc. (PB-264 543) and by Engineering Extension Service, Auburn University (PB-265 713).

NO_x REDUCTION TECHNIQUES

Oxygen Level Controls

The principal method described here for improving boiler efficiency and reducing NO_x emissions involves operating the boiler at the lowest practical excess O_2 level. These O_2 levels will be at some operating margin above the absolute "minimum O_2" which is at the threshold of smoke or combustible emissions formation. Although peak boiler efficiency will occur close to the minimum O_2, it is not practical to operate at this condition unless the boiler is equipped with highly sophisticated combustion controls and flame quality monitoring to prevent any small deviations into unsafe or unacceptable combustible conditions.

Since these control features are not typical of most industrial boilers, some margin or operating "cushion" above the minimum O_2 will be necessary to accommodate normal variations in fuel properties and atmospheric conditions, repeatability and response characteristics of the combustion control system, and other operational factors.

To determine the lowest practical O_2 levels for a particular boiler requires that the minimum O_2 and the appropriate operating margin above this minimum O_2 be established and evaluted with respect to typical levels for the type of fuel and burning equipment used. If minimum O_2 levels are found to be excessive, then burner adjustments are recommended as a possible means for reducing the burner's minimum O_2 requirement. High minimum O_2 can also result from improper maintenance of burner equipment (plugged orifice, broken diffusers, etc.) but these problems can be minimized by performing a preliminary boiler inspection.

The advantage of operating with low excess oxygen from the standpoint of efficiency was described in the second chapter. It was shown that with lower excess O_2, the "dry stack gas losses" are minimized leading to highest efficiencies.

The advantages of low excess O_2 with respect to NO_X emissions are illustrated in Figure 5.1. The curves in the three graphs show how NO_X emissions changed as the excess O_2 was varied on a variety of gas-, oil- and coal-fired boilers. It is immediately apparent that NO_X emissions can vary significantly from boiler to boiler even when compared at the same excess O_2. For the most part, NO_X emissions were reduced as the excess O_2 was lowered.

Coal fired boilers had the highest NO_X emission levels and the NO_X was generally found to be more sensitive to changes in excess oxygen than the average oil or gas fired units. The fuel nitrogen content of coal is the major factor in these higher emissions. Industrial boilers fired with natural gas showed the lowest sensitivity of NO_X to excess oxygen changes. In some cases, NO_X actually increased as excess O_2 was lowered. When this occurs, it is still desirable to operate at low excess O_2 for high efficiency but attempts should be made to reduce NO_X emissions by burner adjustments.

Setting up a boiler for low excess O_2 firing will be accomplished through a systematic, organized series of tests. Following a test that documents "as found conditions," the lowest possible level of excess oxygen for the boiler will be established. The lowest level should be found at several firing rates within the boiler's operating range. The actual number of firing rates tested will depend on the design of the boiler control system. Enough firing rates should be tested to assure that after the final control adjustments are made the optimum excess O_2 conditions are maintained at all the intermediate firing rates.

FIGURE 5.1: EFFECT OF EXCESS OXYGEN ON NITROGEN OXIDES EMISSIONS

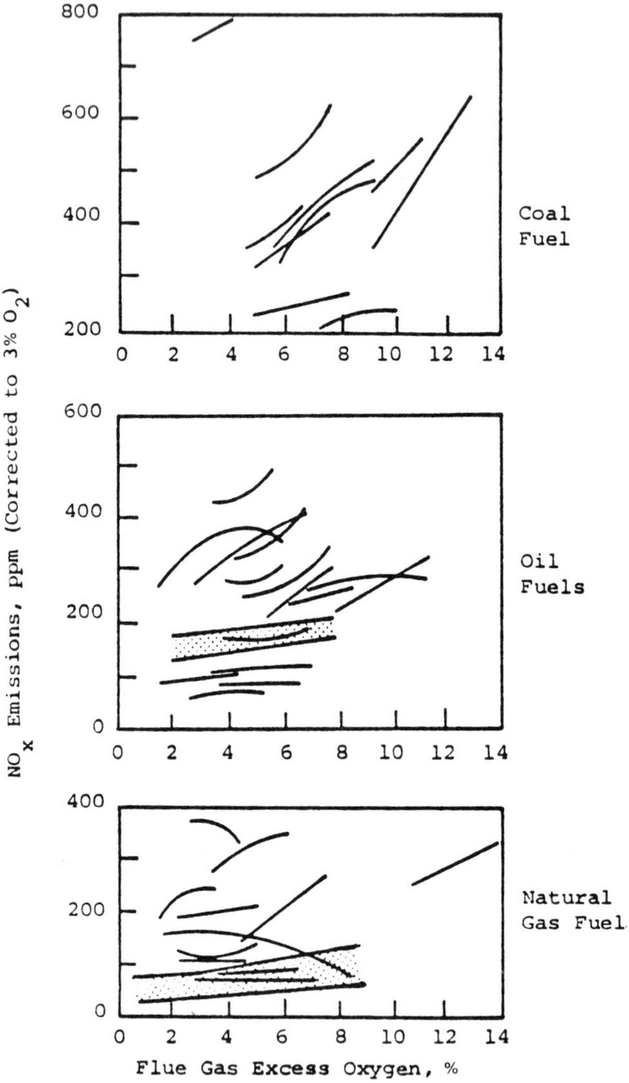

Single lines for watertube boilers; shaded areas represent all firetube boiler data.

Source: PB-264 543

At each firing rate tested, the excess O_2 should be varied over a range from 1 or 2% above the normal operating point down to the point where the boiler just starts to smoke or the CO emissions rise above 400 ppm. This low excess O_2 condition is referred to as the smoke or CO threshold (limit) or simply the "minimum O_2." The smoke limit applies to coal and oil firing since smoking will generally occur before CO emissions reach significant levels. The CO limit applies to gas fuels and is the lowest possible excess O_2 level while maintaining carbon monoxide below 400 ppm.

The smoke limit for coal or oil fuels is the lowest possible excess O_2 level where acceptable stack conditions can still be maintained. In terms of the Smoke Spot Number (SSN) scale, the maximum desirable smoke number on coal fuel for acceptable stack conditions is SSN 4 (measured downstream of the dust collector). For fuel oils, the maximum smoke levels are given in Table 5.1 (1).

TABLE 5.1: MAXIMUM SMOKE LEVELS

Fuel Grade	Maximum Desirable SSN
No. 2	<1
No. 4	2
No. 5*	3
No. 6	4

*(Light and heavy), and low-sulfur resid.

Source: PB-264 543

When performing the boiler tests to determine minimum excess O_2, curves such as those in Figures 5.2 and 5.3 will be constructed. Based on measurements obtained during the tests, these curves will show how the boiler smoke and CO levels change as the excess O_2 is varied. Each of the figures contains two distinct curves to illustrate the extremes in smoke and CO behavior which may be encountered. The curves labeled (**1**) exhibit a very gradual increase in CO or smoke as the minimum O_2 condition is reached.

In contrast, the curves labeled (**2**) are gradual at first but as the excess O_2 is reduced further and minimum O_2 is approached, the smoke or CO emission increases rapidly. In this case, unpredictably high levels of smoke or CO or potentially unstable conditions can occur with small changes in excess O_2 and extreme caution is required. When making O_2 changes near the smoke or CO limit, it should be done in very small steps until there are enough data to show whether the boiler has a gradual or steep characteristic curve for smoke and CO [curve (**1**) versus curve (**2**)]. It is important to note that the boiler may exhibit a gradual smoke or CO behavior at one firing rate and a steep behavior at another.

High minimum O_2 may be an indication of a burner malfunction or other fuel or equipment related problems. But it should also be realized that different burner designs and fuels will in general have different minimum O_2 requirements. Many burners will also exhibit higher minimum O_2 at lower firing rates. For these reasons it is difficult to specify a range of minimum O_2 levels which would be considered normal. However, based on numerous minimum O_2 tests at industrial boiler installations, Table 5.2 is helpful in judging whether the minimum O_2 levels that are measured are typical values.

FIGURE 5.2: TYPICAL SMOKE-O_2 CHARACTERISTIC CURVES FOR COAL OR OIL-FIRED INDUSTRIAL BOILERS

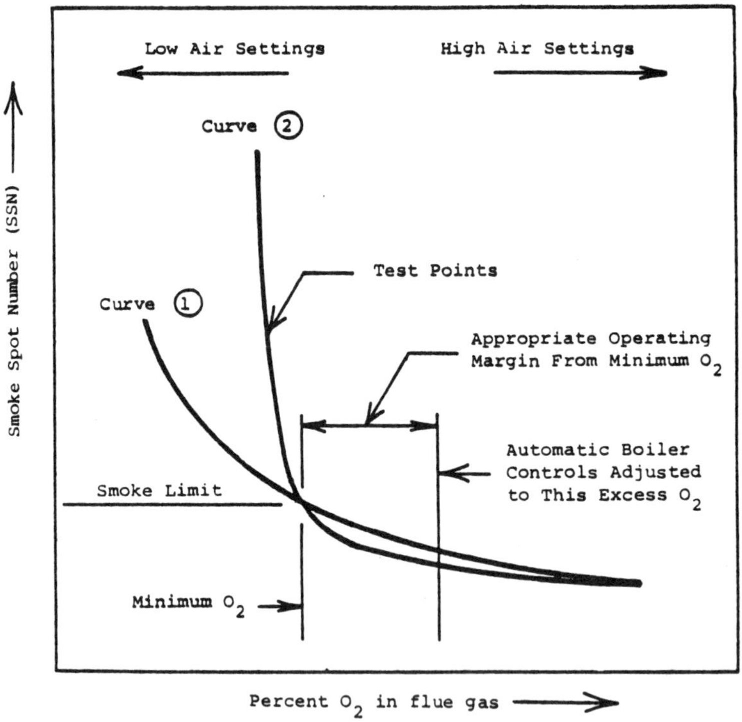

Curve 1 - Gradual smoke/O_2 characteristic

Curve 2 - Steep smoke/O_2 characteristic

Source: PB-264 543

FIGURE 5.3: TYPICAL CO-O_2 CHARACTERISTIC CURVES FOR GAS-FIRED INDUSTRIAL BOILERS

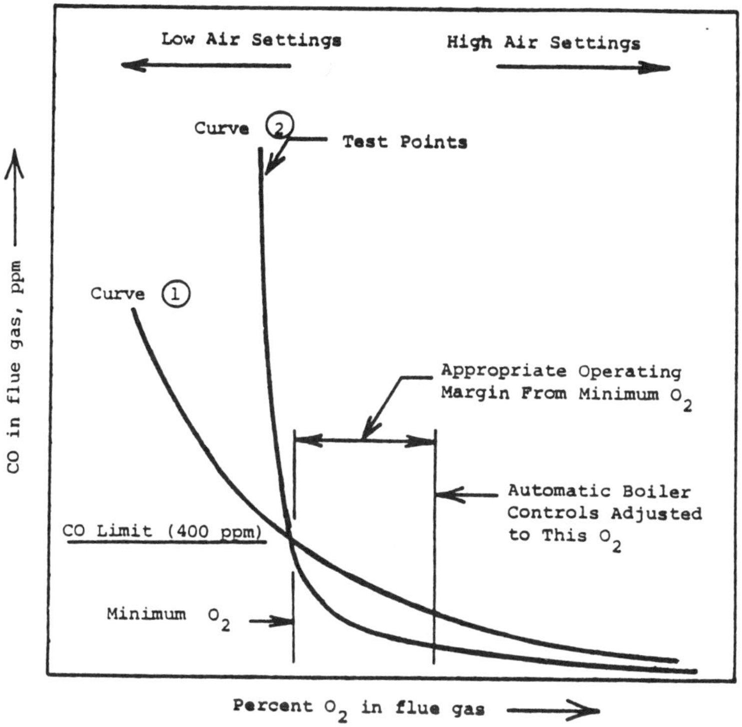

Curve 1 - Gradual CO/O_2 characteristic

Curve 2 - Steep CO/O_2 characteristic

Source: PB-264 543

TABLE 5.2: TYPICAL MINIMUM O_2 LEVELS

Fuel Type	Typical Range of Minimum Excess O_2 At High Firing Rates, %
Natural gas	0.5-3.0
Oil fuels	2.0-4.0
Pulverized coal	3.0-6.0
Coal stoker	4.0-8.0

Source: PB-264 543

Boiler start-up records showing initial excess O_2 conditions can provide a valuable comparison with current minimum O_2 conditions when they are available. If it is apparent that current minimum O_2 levels are higher than the expected levels, there may be a need for maintenance or repairs which should be completed before attempting the following excess O_2 optimization procedure.

Once the minimum O_2 is established, the next step will be to determine the appropriate O_2 margin or operating "cushion" above the minimum O_2 where the boiler can be routinely operated. This will be the lowest practical O_2 for the boiler and will be the optimum setting for high efficiency (and in most cases, lowest NO_X emissions). The O_2 margin above the minimum O_2 is necessary to account for uncontrollable variations in the excess O_2 resulting from:

1) Rapid boiler modulation which could result in smoking or combustibles without an adequate O_2 margin.
2) Nonrepeatability or "play" in the automatic controls (excessive play should be corrected).
3) Normal variations in atmospheric conditions which can also change excess O_2 on units not equipped with temperature and pressure-compensated combustion air systems. This is a very important factor which is often neglected. (Extreme variations in ambient conditions can easily produce changes in excess O_2 of 1% or more.)
4) Changes in fuel properties which may require varying amounts of excess air.

Typical O_2 margins above the minimum O_2 may range from 0.5 up to 2.0% O_2, depending on the characteristics of the particular boiler control system and fuels. When in doubt, the boiler manual or manufacturer should be consulted. When the boiler is going to be operated at a constant firing rate for extended periods, the lowest possible O_2 margin should be selected.

A simple test to determine whether there is excessive "play" (or poor repeatability) in the boiler controls can be made by simply repeating a firing rate condition, allowing the fuel and air controls to function in a normal manner. When performing this test, the particular firing rate should be approached from both the "high side" and "low side," (i.e., from higher and lower firing rates). A comparison of stack excess O_2

after the boiler has stabilized indicates whether there is excessive wear or tolerance problems in air dampers, control shafts, valve cams, controllers, etc. Excess O_2 should repeat to within a few tenths percent excess O_2.

Atmospheric variations are reflected in excess O_2 variations due to changes in air density at the forced draft fan inlet as the air pressure and temperature change. The fan will deliver a certain volume of air to the burner, but as the air density changes, the pounds of air supplied to the burner will vary, thus changing the burner excess air and excess O_2. When combustion air is supplied from the boiler room, maintaining constant boiler room temperatures will minimize this problem, but atmospheric pressure effects are unavoidable.

ALTERNATE NO_x REDUCTION TECHNIQUES

As just described operation of most industrial boilers with low excess air will improve boiler efficiency as well as reduce NO_x stack emissions. Where further NO_x reductions are required or when efforts to improve efficiency lead to increases in NO_x, there are several other NO_x reduction techniques that may be considered. When properly applied, they will usually involve little or no sacrifice in boiler efficiency (except for the reduced air preheat method). These techniques include:

1) Fuel-rich (staged) combustion
2) Flue gas recirculation
3) Low NO_x burners
4) Change of fuels
5) Reduced air preheat

In most cases, applying any of these methods will involve modifications to existing hardware and boiler operating practices. The selection of the appropriate method will require the assistance of outside combustion specialists. Economic evaluations including cost/benefit studies and long-term maintenance and operating costs will also be involved in selecting the most appropriate technique. Whether or not a specific approach can be successfully applied will generally depend on the design and particular operating requirements of the boiler.

Before any of these techniques is considered, the reduced-NO_x potential of existing equipment should be fully evaluated. This includes minimizing the burner excess air level and exploring possible burner adjustments.

Fuel-Rich Combustion

Fuel-rich or "staged" combustion is mainly applicable to gas, oil and pulverized coal fired boilers with multiple burners (four or more burners). This technique usually involves terminating or reducing the fuel to one or more of the burners and equally distributing this fuel to the remaining burners to maintain the same overall firing rate. Full air flow is maintained at the burners with reduced or terminated fuel flow thereby causing the other burners to operate more fuel-rich.

In this manner, the fuel-rich burners can actually be operated with less air than required for complete combustion. This requires the flame products to be effectively mixed with "make up" air from the out-of-service or reduced-fuel burners to complete the combustion elsewhere in the furnace. The fuel-rich flames provide an opportunity for a large portion of the fuel to burn in an O_2 deficient and relatively "cool" flame which reduces NO_X formation. When the flame products eventually mix with the make up air downstream in the furnace, the combustion gas temperature is already too low for additional NO_X formation but adequate to complete the burnout of smoke and combustibles.

An alternate approach involves maintaining normal fuel flow to all burners while directing a portion of the combustion air through auxiliary or "over fire" air ports located away from the burners. However, this requires redesign of the air handling parts of the boiler and therefore can be costly.

NO_X emission reductions of 25 to 40% have been demonstrated with fuel-rich firing (2) but it is seldom a straightforward technique in terms of implementation by the boiler owner/operator. It usually requires an extensive test program to establish the most effective fuel distribution pattern and to addresss potential problems such as reduced load capacity, flame stability, furnace heat release effects, flame impingement, and tube wastage. This work is best performed by combustion specialists with previous experience in fuel-rich firing techniques. However, this approach is relatively cost effective compared to other NO_X reduction techniques in that little or no hardware modifications are required with the burners out-of-service method of fuel-rich firing.

Where this technique is applicable, it is very attractive since it can involve little or no decrease in boiler efficiency for large reductions in NO_X. Efficiency penalties mainly arise when boiler excess air requirements increase to maintain acceptable stack conditions or combustible losses. In some cases the excess O_2 requirement will increase only a few tenths percent and the resulting efficiency penalty is extremely small.

Flue Gas Recirculation

A very effective means of reducing nitrogen oxides emissions involves recirculating a portion of the flue gas back into the combustion air. These recirculated gases absorb heat energy from the flame thereby lowering the peak flame temperature which reduces NO_X formation. Other effects on NO_X may result from increased flame turbulence due to higher mass flow through the burner and alteration of flame shape and heat release.

Adding a flue gas recirculation system can conceivably be done on most boilers but it usually requires considerable costs, engineering design and extensive equipment modification. The point where the flue gas is introduced into the combustion air must be carefully chosen and designed to assure thorough mixing of flue gas and air before it flows through the burners. Changes in burner design are sometimes necessary to retain flame stability.

Flue gas recirculation has been demonstrated to be generally more effective in reducing NO_X emissions from gas fired boilers (3). Reductions in NO_X of up to 73% at gas fueled industrial boilers have been achieved. Reductions with fuel oil can be on the order of 30%. Flue gas recirculation is less effective with heavy oil and coal fuel because they contain high levels of fuel nitrogen. The formation of NO_X from fuel nitrogen is not strongly influenced by gas recirculation. Flue gas recirculation can affect overall plant efficiency slightly due primarily to increased auxiliary power requirements (gas recirculation fans). A reduction in boiler excess O_2 is sometimes possible using flue gas recirculation but the increased auxiliary loads usually override this beneficial effect on boiler efficiency.

Low-NO_X Burners

Retrofit burners are commercially available which are specially designed for low nitrogen oxides emissions. These "low-NO_X" burners give a controlled slow-burning, low-temperature flame. Some designs involve low excess O_2 operation which is also beneficial from a boiler efficiency standpoint. Interpreting low-NO_X burner test data from other boilers must be done carefully before considering the use of the burner at another installation. Many factors such as fuel composition, furnace geometry and combustion control precision are influential in the success of a low-NO_x burner.

For instance, a low-NO_x burner design which works well on a single burner installation may not give beneficial results when used in a multiple burner unit, and vice versa. Careful selection by engineering personnel assisted by competent combustion specialists is necessary in choosing a low-NO_x burner best suited for any particular boiler. It is sometimes possible that an existing burner is capable of producing NO_x emission levels of a "low-NO_x" burner if properly maintained and adjusted.

Fuel Change

Changing fuels can be an effective method for reducing oxides of nitrogen. This might involve changing from a higher nitrogen fuel oil to a lower nitrogen oil (i.e., No. 6 to No. 2) or by selecting oils of the same grade which have lower nitrogen content. For example, variations in the fuel nitrogen content of No. 6 oils of more than a factor of 3 can occur as a result of refining processes, blending and natural differences between oil fields.

Changing fuel types (i.e., from coal to oil or oil to gas) is considerably more complicated and the decision is usually controlled by other factors such as fuel availability and cost. Also it is not always practical because of physical limitations of the boiler. A boiler may require extensive burner modifications, new fuel handling equipment and increased soot blowing and ash handling facilities to burn a different fuel. Fuels such as natural gas and light distillate oils which potentially produce the lowest NO_x emissions may be expensive or in increasingly short supply. Different fuels characteristically exhibit modest differences in boiler efficiency and this factor should be taken into account in addition to raw fuel cost (on a million Btu basis) when comparing alternate fuels. These differences result primarily from dissimilar hydrogen content in the fuels with resulting differences in moisture content of the flue gas (moisture losses).

Reduced Combustion Air Preheat

Many larger industrial boilers operate with preheated combustion air to increase boiler thermal efficiency and improve combustion. Unfortunately, heated combustion air results in higher flame temperatures which promotes higher nitrogen oxides emissions. Combustion air preheaters are heat exchangers which extract heat from the flue gas to heat the incoming combustion air. Some preheaters are equipped with ducting for bypassing a portion of the air or flue gas around the preheater to control stack temperatures for corrosion purposes. This capability can also be used in some cases to lower combustion air preheat and effectively reduce NO_X emissions. The major drawback with this technique is that boiler efficiency is reduced as reflected by higher stack temperature (higher stack gas losses). The trade-off between efficiency and NO_X emissions is the major consideration.

This technique is most effective on natural gas and light fuel oils; however, reductions are possible with heavy oils and coal. Potential operational problem areas include flame stability and possible degradation in combustion conditions leading to increased furnace tube fouling.

TREATMENT OF BOILER FEEDWATER

Water quality is a problem that must be considered. Untreated or raw feedwater can cause numerous problems in boiler operation, such as scale formation, corrosion, foaming, etc. Thus it is necessary to keep a watchful eye on boiler feedwater quality to ensure clean and efficient boiler operation.

Water Quality Terminology

Only those terms pertaining to water quality terminology which are likely to be used in describing boiler feedwater quality are used here.

Total hardness is the amount of hardness in the water due to calcium and magnesium salts.

Alkalinity is the amount of bicarbonate, carbonate, and hydroxide (or caustic) compounds in the water. Alkalinity is usually measured in pH, a measure of the concentration of acids and/or basic compounds.

Silica is the soluble silica (SiO_2) present in the water. Silica scale is highly resistant to heat transfer and, therefore, is very detrimental to boiler feedwater.

Total solids is the amount of dissolved solids and particulate matter in the water. Dissolved solids refers to the amount of dissolved salts and minerals while particulate solids refers to the soil and biological matter present in the water.

A list of typical feedwater contaminants is given in Table 5.3.

Types of Feedwater Treatments

Feedwater treatments may be broken down into three major categories, Chemical, Thermomechanical, and Electrochemical. Within each of these categories there are two distinct types of treatments: internal and external. Internal treatment refers to those processes that take place in the boiler, while external refers to those processes that take place before the feedwater enters the boiler.

Chemical Treatments: *Ion-Exchanger Absorption*—This process is classified as external and requires a separate reactor vessel equipped with a surface-type chemical bed. Essentially, the bed exchanges undesirable ions in the feedwater with less objectionable ions in the bed. The ion-exchanger is rejuvenated by backwashing and adding a chemical that reverses the process, releasing the previously captured impurity ions and replacing them with the desirable process ions. This reaction is generally used to replace insoluble compounds with soluble compounds that may require further processing.

Chelating Agents—Chelating agents are classed as internal, as they are added to the feedwater as it enters the boiler. The chelating agent combines with chemically unstable impurities (those most likely to form scale) to form stable soluble compounds.

Precipitating Agents—Precipitating agents may be used either externally or internally. The precipitating agent combines with impurities to form insoluble compounds that settle out and are exhausted as sludge.

Chemical Degasifiers—Chemical degasifiers are used internally to remove any dissolved gasses that may have passed external feedwater treatment. The chemical agent reacts with oxygen and corrosive soluble salts present to form chemically stable (noncorrosive) soluble compounds.

Coagulating Agents—Coagulating agents are used external to the boiler. Coagulating agents are used with other treatments to form a semisolid aggregate of impurities that can be removed by a mechanical filter.

TABLE 5.3: COMMON IMPURITIES IN WATER AND POSSIBLE EFFECT WHEN USED FOR BOILER FEED PURPOSES

Constituent	Chemical Composition	Principal Source of Contaminating Substance	Possible Effect when Present in Boiler Waters
Calcium bicarbonate	$Ca(HCO_3)_2$	Mineral deposits	Scale
Calcium carbonate	$CaCO_3$	Mineral deposits	Scale
Calcium chloride	$CaCl_2$	Mineral deposits	Scale
Calcium sulfate	$CaSO_4$	Mineral deposits	Scale and corrosion
Carbonic acid	H_2CO_3	Absorption from the atmosphere, mineral deposits, decomposition of organic matter	Corrosion
Free acids	HCl, H_2SO_4	Mine drainage and industrial wastes	Corrosion
Magnesium bicarbonate	$Mg(HCO_3)_2$	Mineral deposits	Scale
Magnesium carbonate	$MgCO_3$	Mineral deposits	Scale
Magnesium chloride	$MgCl_2$	Mineral deposits	Scale and corrosion
Oil and grease	–	Industrial wastes	Corrosion, deposits, priming and foaming
Organic matter and sewage	–	Domestic and industrial wastes	Corrosion, deposits, priming and foaming
Oxygen	O_2	From atmosphere	Corrosion
Silica	SiO_2	Mineral deposits	Scale
Sodium bicarbonate	$NaHCO_3$	Mineral deposits	Priming, foaming and embrittlement
Sodium carbonate	Na_2CO_3	Mineral deposits	Priming, foaming and embrittlement
Sodium chloride	NaCl	Sewage, industrial wastes, and mineral deposits	Inert, but may be corrosive under some conditions
Suspended solids	–	Surface drainage and industrial wastes	Priming, foaming, sludge or scale

Source: PB-265 713

Thermomechanical Treatments: *Degasifiers*—Most degasifiers are used outside the boiler; they heat the feedwater to near boiling and then vent the excess gases. Typically this operation is about 97% efficient.

Filters—Mechanical filters are externally used in most boiler systems. Filters are generally used to remove particulate matter and coagulating agents. Some filters use the mechanical properties of the water for separation and others simply screen out particles greater than a prescribed size.

Electrochemical Treatments: *Cathodic Reactors*—The cathodic reactor (or protector) is an internal device that causes an electric field to collect metal ions. At the anode a suitable material is used so that the anode reacts with the collected impurities to produce chemically stable soluble compounds and some precipitating compounds that are exhausted during blowdown.

Contaminant Analyses

Since there are a large number of water contaminants it is necessary to sort out specific contaminants that cause the greatest trouble in boilers. The contaminants that most affect boiler feedwater quality are given in Table 5.4.

TABLE 5.4: CONTAMINANT EFFECTS

What to Test for	What it Affects
Dissolved solids Dissolved metals Undissolved solids	Total solids
Calcium hardness Magnesium hardness	Total hardness
Biocarbonates Carbonates Caustics (hydroxides)	Alkalinity
Dissolved O_2 Dissolved CO_2	Dissolved gases
Silica	Silica

Source: PB-265 713

Water quality may be tested in one of several ways. Most state and local governments operate water quality laboratories, and some private institutions also have laboratories. A water sample may be obtained and sent to one of these laboratories.

Sample recommended limits of impurities are shown in Table 5.5.

TABLE 5.5: RECOMMENDED IMPURITY LEVELS

Contaminant	Recommended Level (concentration)
Total dissolved solids	0.1 ppm
Suspended solids	zero
Hardness	zero
Free caustic	zero
Dissolved O_2	<0.007 ppm
Dissolved CO_2	next to zero
Silica	0.01 ppm
Iron	0.01 ppm
Copper	0.005 ppm
pH	8.5-9.5 pH at 77°F

Source: PB-265 713

No single process is satisfactory for all contaminants; several processes are required to reduce the impurities to allowable levels. Table 5.6 shows the boiler problem, its probable source and the processes that have been adapted to remove the impurities.

TABLE 5.6: WATER QUALITY TROUBLESHOOTING CHART

Boiler Problem	Cause	Applicable Processes
Scale	Total hardness Silica content	Softening by heaters and chemicals Ion-exchange process Evaporation Coagulating agents Precipitating agents Cathodic reactors Chelating agents
Corrosion	Dissolved gases Alkalinity	Degasification by thermomechanical or chemical means Cathodic reactors Ion-exchange process Evaporation Precipitating agents Chelating agents

(continued)

TABLE 5.6: (continued)

Boiler Problem	Cause	Applicable Processes
Sludge	Undissolved solids	Coagulating agents Precipitating agents Mechanical filters Chelating agents Cathodic reactors Evaporation
Embrittlement	Alkalinity	Chemical softening Ion-exchange process Evaporation Chelating agents Precipitating agents Cathodic reactors

Source: PB-265 713

REFERENCES

(1) *Guidelines for Burner Adjustments of Commercial Oil-Fired Boilers,* EPA-600/2-75-069-b.

(2) *Field Testing: Application of Combustion Modifications to Control Pollutant Emissions from Industrial Boilers - Phase II,* EPA-600/2-76-086a, April 1976.

Operational Procedures and Modifications

The material in this chapter was excerpted from reports by KVB, Inc. (PB-262 577), the Engineering Extension Service, Auburn University (PB-265 713) and the Federal Power Commission (TID-27600).

BOILER OPERATIONAL MODIFICATIONS

This section outlines some changes in boiler operating practices that can be employed to reduce fuel consumption. These include not only techniques that specifically improve boiler efficiency but also fuel saving practices such as load management and reduced continuous blowdown practices.

Reduced Boiler Steam Pressures

Reducing the boiler steam pressure is not generally viewed as an efficiency improvement technique but at those boiler installations where reductions in steam pressure are practical, this can be an effective means for saving up to 1 or 2% on the fuel bill. A portion of this savings results from lower stack temperatures and the accompanying increase in boiler efficiency. Lower steam pressures give lower saturated steam temperatures and in cases without stack heat recovery, a similar reduction in final flue gas temperature will result. The actual reduction in temperature for a given drop in steam pressure will depend on the particular pressure levels involved, since the relationship between saturated steam pressure and steam temperature is not linear.

For example, a 50 psi drop from 150 psig to 100 psig will lower the steam temperature (and stack temperature) by 28°F. By comparison, a 50 psi drop from 400 psig will reduce the stack temperature by 12°F. The corresponding boiler efficiency improvements would be approximately 0.7% and 0.3% respectively.

There can be other immediate benefits from operating at reduced pressures which lead to lower fuel consumption. These can include lower radiated heat loss from steam mains, less leakage at flanges and packing glands, reduced boiler-feed-pump energy consumption and less energy dissipation at pressure reducing stations.

To successfully operate at reduced pressures, the primary pressure must still be sufficiently high to deliver adequate steam to the various points of end use. Lower pressures may also result in undesirable moisture carryover from the steam drum. The boiler manufacturer may be able to supply improved drum intervals if carryover is a problem. While these are usually the primary concerns, there may be other important considerations.

For example, operating with lower main steam pressure might place tighter limits on acceptable steam pressure fluctuations. Since pressure drops across regulators are less, smaller transients in input pressure are acceptable to assure uniform output. Flow metering devices may also have to be recalibrated or corrected for the lower pressures.

Water Quality Control – Blowdown

Water treatment is an important aspect of boiler operation which can affect efficiency or result in plant damage if neglected. Boiler feedwater contains impurities in solution and suspension. These impurities concentrate in the boiler water since the steam generated is essentially pure. Proper internal boiler water treatment removes all or part of the scale-forming calcium and magnesium salts from solution. If these suspended solids are allowed to concentrate beyond certain limits, a deposit or "scale" will form on the boiler heating surfaces which will retard heat transfer and increase tube metal temperatures. This can lead to increased stack gas temperatures which reduces boiler efficiencies.

Even more important may be the probability of furnace tube failures from overheating as a result of the insulating effect of water-side scale. High solids can also interfere with proper operation of the steam separating apparatus in the drum causing boiler water carryover with the steam. If a superheater is installed, the solids will either be deposited on the surfaces causing metal overheating or be carried along with the superheated steam. Particulates in the steam can cause further problems in erosion or deposits in turbines and valves.

The concentration of suspended and dissolved solids in boiler water is controlled by removing some of the high-solids boiler water and replacing it with low-solids feedwater, effecting a general lowering of solids concentration in the boiler. This process, which is known as blowdown, can be either intermittent bottom blowdown or continuous blowdown. Bottom or sludge blowdown is necessary to remove any sludge accumulating in the lowest parts of the boiler system. Continuous blowdown is taken from the point of highest solids concentration usually from the upper drum.

Use of continuous rather than intermittent blowdown alone saves treated boiler water and can result in significant energy savings. Typical intermittent blowdown patterns shown in Figure 6.1 can result in periodic shifts in total solids concentration and require higher blowdown rates to avoid solids carryover with the steam. Actual fuel savings in changing over to continuous blowdown depends on the existing blowdown patterns. The use of continuous blowdown will also permit the recovery of blowdown heat energy.

Typical energy savings that can be generated by recovering heat in blowdown water can be estimated using Figures 6.2 and 6.3. The percentage of blowdown will be fixed by the solids concentration of the make-up water and the maximum allowable boiler water.

Manufacturer estimates of fuel savings using a semipressurized and fully pressurized condensate return systems as compared to open systems are approximately 12 to 15% respectively. Actual savings will depend on boiler operating pressure, flash tank pressure and the eventual use of the flashed steam. Preheating boiler feedwater with waste heat can amount to a 1% savings for every 10°F increase in feedwater temperature.

Many industrial applications involve a high percentage of make-up water with varying amounts of impurities. Blowdown rate, which is defined as the ratio of water blown from the boiler to the quantity of boiler feedwater, may range up to 10% depending on the ratio of impurities

FIGURE 6.1: VARIATION OF BOILER WATER CONCENTRATION (Ref. 1)

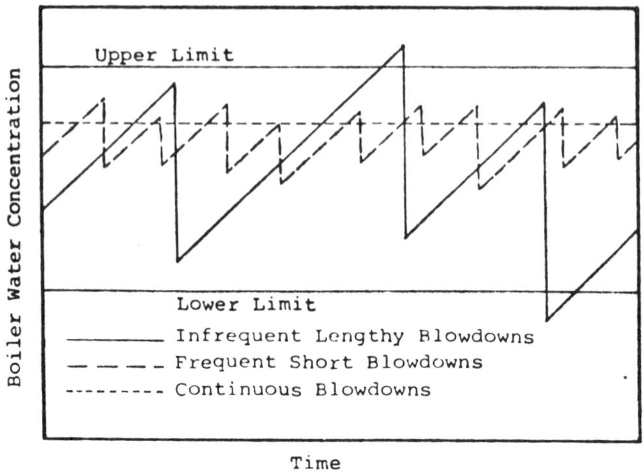

Source: PB-262 577

FIGURE 6.2: PERCENT BLOWDOWN REQUIRED TO MAINTAIN PREDETERMINED BOILER-WATER DISSOLVED SOLIDS CONCENTRATION (Ref. 2)

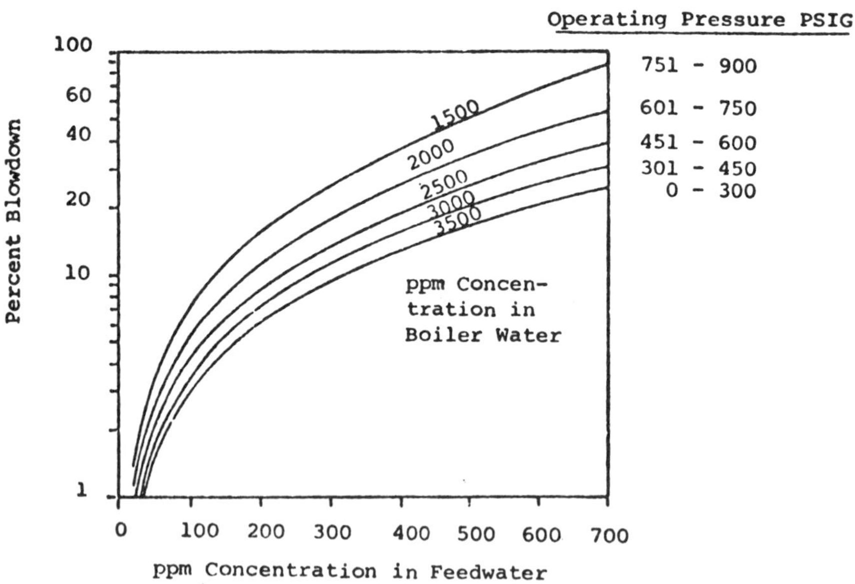

Source: PB-262 577

FIGURE 6.3: WASTE HEAT ENERGY RECOVERY POTENTIAL

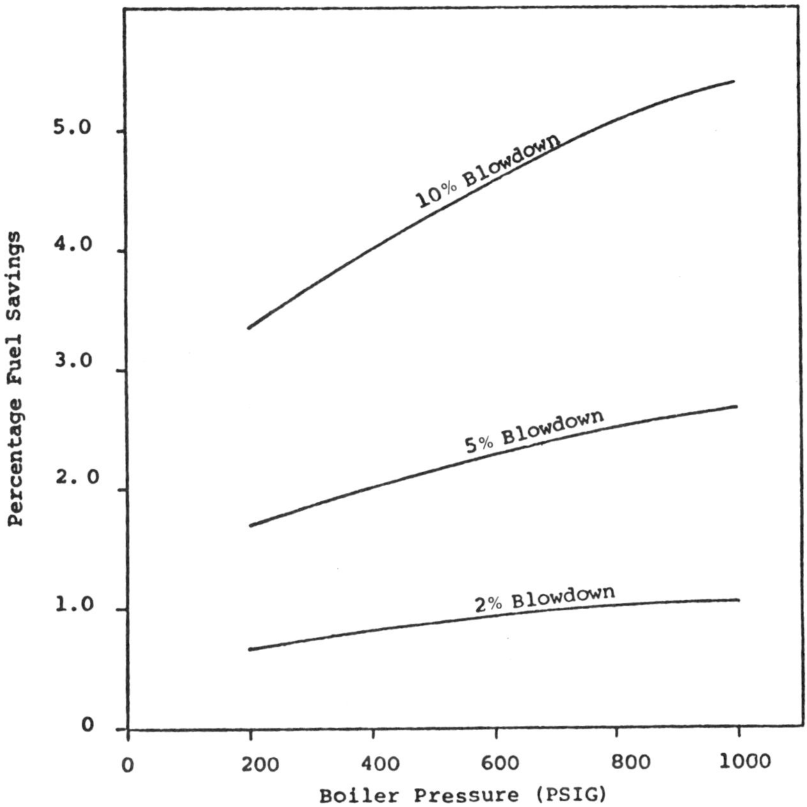

Source: PB-262 577

in the feedwater to allowable concentration in the boiler water. For instance, the permissible concentration of total solids in boiler water is 3,500 ppm at 300 psi or lower operating pressure. If the feedwater contains 350 ppm of total solids, a blowdown rate of 10% would be required to maintain the solids level in the boiler at 3,500 ppm. The heat loss associated with 10% blowdown is roughly 3% if none of the sensible heat carried away in the blowdown water is recovered.

Nearly all of this heat can be recovered in a continuous blowdown system incorporating heat exchangers which cool the blowdown to within 20° to 30°F of incoming make-up water as will be discussed. The potential saving is proportional to the amount of blowdown required which further depends on the percentage and purity of make-up water. This is discussed in more detail in the chapter on Removing Impurities. Preboiler treatment of water to reduce solids can also reduce blowdown requirements. A firm specializing in industrial water treatment should be consulted to determine if preboiler water treatment would provide savings.

Physical Description: Table 6.1 presents the recommended maximum concentration of impurities that can be tolerated in the boiler above which priming, foaming, and carryover occur resulting in low quality steam. As mentioned, excessive concentrations of dissolved and suspended solids also lead to internal scale formation, reduced heat transfer and, ultimately, tube failures. Therefore blowdown systems remove boiler water with high impurity concentrations and replace it with clean make-up water. Make-up water deionizers can also be employed to remove dissolved minerals thus reducing blowdown requirements.

TABLE 6.1: RECOMMENDED MAXIMUM IMPURITY CONCENTRATION

	Boiler feedwater			Boiler water		
Drum pressure, psig	Iron, ppm Fe	Copper, ppm Cu	Total hardness, ppm $CaCO_3$	Silica, ppm SiO_2	Total alkalinity, ppm $CaCO_3$	Specific conductance, micromhos/cm
0-300	0.100	0.050	0.300	150	700	7000
301-450	0.050	0.025	0.300	90	600	6000
451-600	0.030	0.020	0.200	40	500	5000
601-750	0.025	0.020	0.200	30	400	4000
751-900	0.020	0.015	0.100	20	300	3000
901-1000	0.020	0.015	0.050	8	200	2000
1001-1500	0.010	0.010	ND	2	0	150
1501-2000	0.010	0.010	ND	1	0	100

(Reprinted with permission from *Power* Magazine, February, 1976)

Source: PB-262 577

As just described, significant savings in hot water can be attained using continuous blowdown. In addition, continuous blowdown allows the use of heat recovery devices to preheat the incoming make-up water with waste heat from the blowdown water. It is common practice to recover as much of the heat in the expelled hot water as possible. Low pressure boilers with low blowdown rates usually employ heat exchangers to preheat the incoming make-up water.

Higher capacity units operating at high pressures use a flash system in which a portion of the wastewater is flashed to high purity steam at low pressures, condensed and recycled to the feedwater system. The remaining wastewater having now an even higher concentration of impurities is passed through heat exchangers to preheat make-up water and discarded.

Condensate Return Systems: Condensate recovered from process steam at boiler operating pressure and temperature represents a valuable quantity of clean water and usable energy. The amount of condensate that can be recovered is dependent on the degree of contamination in the condensed water. In certain cases separator or filters can be employed to remove contaminants such as oil or rust material so that the water is suitable for use in the boiler. Water containing a high degree of contaminants is generally discarded.

The recovery of condensate is also limited by the type of system utilized. The techniques employed to recycle the condensate to the boiler can be divided into five categories: atmospheric (open) system, semipressurized (semiclosed) system, fully pressurized (closed) system, vacuum type and deaerators.

A study must be conducted in any retrofit application to assure that the system chosen is compatible with existing boilers and process requirements.

The atmospheric system employs a vented tank downstream of a steam trap or drain in which the hot condensate is received. This is done to reduce condensate temperature to 212°F to allow pumping without cavitation occurring. A portion of the condensate will flash to steam and vent to the atmosphere representing a loss of both energy and water. Treated make-up water is added as required to meet feedwater flow requirements. This is the most common type of return system with the least cost. It is generally compatible with all units especially when large amounts of condensate are returned. Losses may be reduced by proper sizing and application of steam traps and drains.

The semipressurized system employs a condensate receiver that is at an intermediate pressure at least 50 psi below the lowest anticipated steam pressure in the process system. The intermediate pressure permits use of conventional condensate (trap) return systems, while retaining a significantly higher condensate temperature. The receiver pressure is controlled by steam passing through a pressure reducing valve from the main steam header to the receiver. This steam, along with the steam flashed from the condensate, can be employed to heat make-up water or supply low pressure process steam. A steam relief valve is provided with a set point 10 to 20 psi above the pressure reducing valve setting and prevents abnormal pressure surges.

An air vent is provided on the receiving vessel to allow noncondensable gases to escape. The condensate is cooled by the make-up water to a temperature at least 20°F below the saturation temperature to prevent vapor lock in the boiler feedwater pump. This system has the advantage of reduced water and energy losses. Pressure variations must be controlled to a minimum in the receiver in order to prevent boiler feed pump cavitation. Relatively stable pressure can be maintained by the steam pressure reducing valve and the steam relief valve combination—even in the face of irregular condensate flow and system steam trap malfunctions.

An attempt is made to return all the condensate to the boiler in the fully pressurized system. The entire system is operated at the steam process pressure with flow losses overcome by specially designed water pumps. Operation of a fully pressurized system is restricted to units with nearly constant steam loads as sudden heat demands will cause pressure fluctuations resulting in the flashing of condensate and restricted flow through the system. Vents operating automatically are provided to allow trapped air to be released. These severe operating restrictions often limit its application.

The vacuum type system is normally only used on low pressure heating systems where large pressure drops exist due to long condensate lines or steam traps.

Deaerators preheat feedwater and remove noncondensible and corrosive gases using heat and agitation. Heat is supplied by returned condensate or low pressure steam while agitation is provided by sprays or trays. Deaerators normally operate at approximately 15 psig. The small amount (1 to 2%) of flashed steam not condensed with the feedwater is vented with the noncondensible gases and later condensed in a vent condenser.

Costs: Capital costs for a continuous blowdown unit will range from $600 to $2,500, depending on the capacity. Installed costs can be two to three times this amount on retrofit applications depending on required plumbing modifications and installation.

Deionizers are approximately $1,500 and $2,000 per GPM make-up water for capital and installed costs respectively.

Installed costs for condensate return systems are also highly variable and will be determined by the individual application. Capital costs for semipressurized and fully pressurized systems are approximately $3,000 per unit with installation costs as high as $8,000.

Capital costs for a deaerating feedwater heater range from $9,000 to $26,000, depending on size, type and configuration. Installed costs will again be dependent on the specific application.

Load Management

Load management is an effective tool in minimizing fuel consumption in a plant with several boilers. Since boiler efficiency varies with load, boiler design, equipment age, fuel, and other factors, there is an advantage to distributing load to the most efficient boilers and to operating boilers at a load close to their peak efficiency.

In general, a peak operating efficiency will occur at a load less than 100% due to the complex interaction between the stack gas temperature, excess air flow and surface radiation losses. Units with high excess air requirements or significant radiation losses at low loads will have a peak efficiency at a high load. Boilers with constant excess air levels and small radiation losses over the load range will exhibit a peak efficiency at a lower load. Several typical efficiency curves are presented in Figure 6.4. It is usually necessary to determine the efficiency variation with load for each individual unit by measurement.

Proper load management would then require the use of the optimum operating range for each boiler in the system whenever practical and maintaining this load to minimize efficiency losses during load variations.

Plants with several boilers should be managed to achieve optimum system performance. The most efficient boilers should be loaded first to the desired operating level and so on to the least efficient boiler. Conversely, the least efficient boilers should be removed from service first. There also is merit to shutting some boilers down in swing seasons while operating process boilers at peak efficiency as opposed to operating all boilers at reduced load and performance. Another variation in multiple boiler plants would be to schedule one boiler for a specified process, another for space heating and so on. A detailed cost analysis including additional operating personnel and maintenance must be conducted to determine the net overall cost savings in multiple boiler systems.

FIGURE 6.4: TYPICAL BOILER EFFICIENCY vs LOAD DEPENDENCE

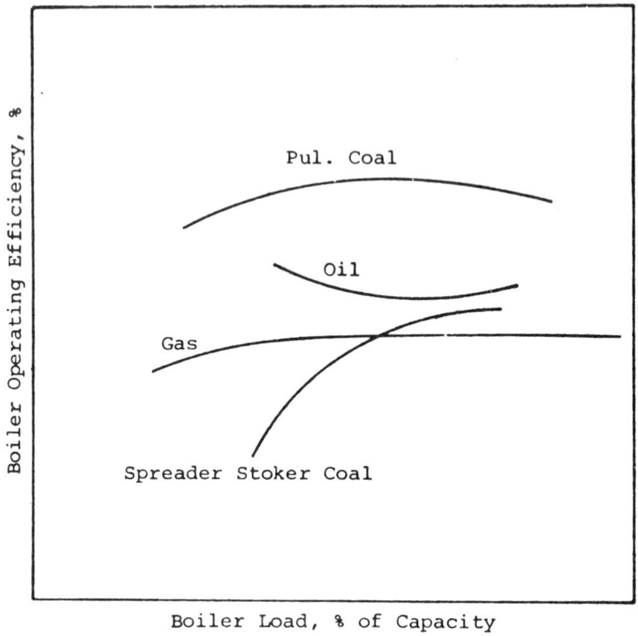

Source: PB-262 577

Proper load management also involves the use of reduced operating levels or shutdown during nonproduction periods. The choice between maintaining the unit at idle position, allowing the unit to be at standby at a reduced firing rate or shutting the unit down is determined by the length of the interruption, the heating and cooling characteristics of the unit, and the effects of the thermal cycles on the life of the unit. Other considerations such as limited turn down range due to fuel firing characteristics or furnace dimensions can also dictate a preliminary shutdown rather than maintaining a low idle firing rate. Reduced operations are always justified from an efficiency standpoint if the energy saved by reduced operation is greater than the energy required to reheat the unit back up to operating conditions.

Selection of new equipment such as quick steaming units that can come to operating temperature and pressures from reduced firing levels in minutes or boilers with maximum efficiency in the range of steam flow utilized may be considered to optimize the load management potential. Several small, more efficient units that meet the specific needs of the process could show a long term economic benefit in terms of fuel savings over fewer larger units.

A thorough engineering evaluation of the cost/benefits and potential effects on equipment life and maintenance must precede the initiation of any load management program.

Fuel Conversions or Change of Fuel

A fuel change or fuel conversion is generally not a common tool for fuel conservation since the decision usually is controlled by other factors such as fuel availability, cost, or process requirements. However, fuel changes can have a substantial impact on the total cost of steam production. Since different fuels do characteristically exhibit modest differences in boiler efficiency, this factor should be taken into account in addition to raw fuel cost (on a million Btu basis). These differences result primarily from dissimilar hydrogen content in the fuels with resulting differences in moisture content of the flue gas (moisture losses), variations in heat release rate in the furnace causing differences in steaming rates and dissimilar slagging and fouling characteristics.

Primary considerations in fuel selection are present and future trends in cost and availability. Other factors include: existing fuel storage and handling equipment, existing combustion equipment and controls, boiler cleanliness problems and sootblowing provisions, air pollution emissions, maintenance, and additional operating personnel and supervision.

A complete evaluation including preliminary testing should be made prior to a long term commitment to an alternate fuel supply.

References

(1) Cato, G.A., et al, *Field Testing: Application of Combustion Modifications to Control Pollutant Emissions from Industrial Boilers - Phase II,* KVB, Inc. Report to the U.S. Environmental Protection Agency, No. EPA 600/2-76-086a, April 1976.

(2) Thompson, R.E., et al. *A Study to Assess the Potential for Energy Conservation through Improved Industrial Boiler Efficiency,* report to the Federal Energy Administration, Contract No. C-04-50085-00, October 1976.

WASTE HEAT RECOVERY

Many existing boilers have considerable potential for waste heat recovery. Described here are general considerations for recovery of waste heat, some processes for recovering this energy, and the possible uses of this energy in an existing boiler system.

There are three main sources of waste heat energy in a boiler system: Exhaust Gases, Process Steam Pressure Drops, and Steam Condensate and Blowdown.

Energy Recovery from Exhaust Gases

Boilers are designed to operate at a particular exhaust gas temperature under specified operating conditions. Boilers operate with different exhaust gas temperatures but they are usually designed to operate with exhaust gas temperatures ranging from 300° to 550°F.

The exhaust gas temperature changes during the operation of a boiler. This temperature is affected by such factors as type of fuel, air-fuel ratio, burner characteristics, fouling of heat transfer surfaces, atmospheric conditions, load, etc.; the exhaust gas temperature can often be reduced by changing one or more of these factors. The exact cause of increases in or high exhaust gas temperatures can be determined by performing tests on the boiler and analyzing the data.

Lowering the exhaust gas temperature will usually increase boiler efficiency. The efficiency should always be measured since some of the factors listed above will lower exhaust gas temperature and also boiler efficiency.

The energy in the exhaust gases may be used to satisfy a number of heating requirements. These requirements may be divided into those associated with the boiler and those associated with other heating needs. For the boiler, the energy in the exhaust gases may be used to preheat water for the boiler, and/or combustion air.

Other heating needs which the exhaust gases may satisfy are heating of buildings, heating of water, and process heating requirements.

The exhaust gases may be used directly to satisfy certain process heating requirements. Most of the uses of this energy requires heat exchangers.

Some heat exchangers which can be used to recover energy from exhaust gas and blowdown of boilers are: heat wheel, finned tube, shell-and-tube, and heat pipe. A brief description of each of these types is given below.

Heat Wheel: The heat wheel is a device that rotates through two streams, hot and colder streams. The hot side of the heat wheel absorbs the energy of the hot gases by raising the sensible heat content of the wheel. The wheel is then rotated to the cold side and a cooler stream of gases is heated by the sensible heat of the wheel. Some heat wheels have been designed to specifically be used in exhaust gases. These wheels are coated with a hygroscopic material that captures part of the latent heat of vaporization of the water vapor in the flue gases as well as the sensible heat of the flue gas (see Figure 6.5).

FIGURE 6.5: HEAT WHEEL

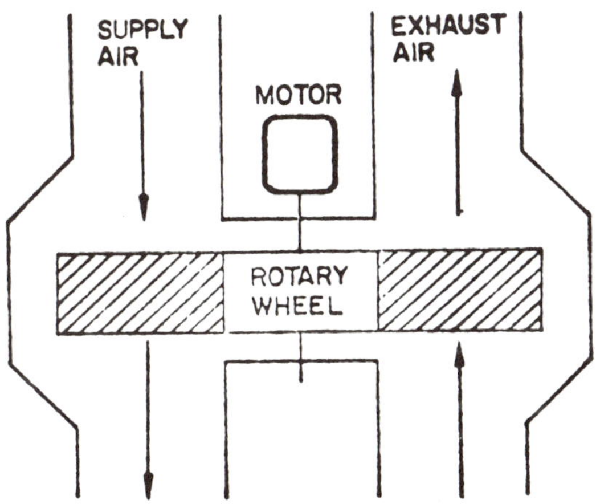

Source: PB-265 713

Heat wheels, in general, are about 85% effective and range in size up to 40,000 cubic feet per minute flow rate and 70 feet in diameter. Heat wheels are often used to preheat combustion air or for hot process air when equipped with a purge section that does not allow cross contamination (see Figure 6.6).

Heat Pipe Heat Exchangers: As the name implies this type of heat exchanger utilizes the heat pipe to transfer energy. Basically the heat

FIGURE 6.6: HEAT WHEEL EQUIPPED WITH PURGE SECTION TO CLEAR CONTAMINANTS FROM THE HEAT TRANSFER SURFACE

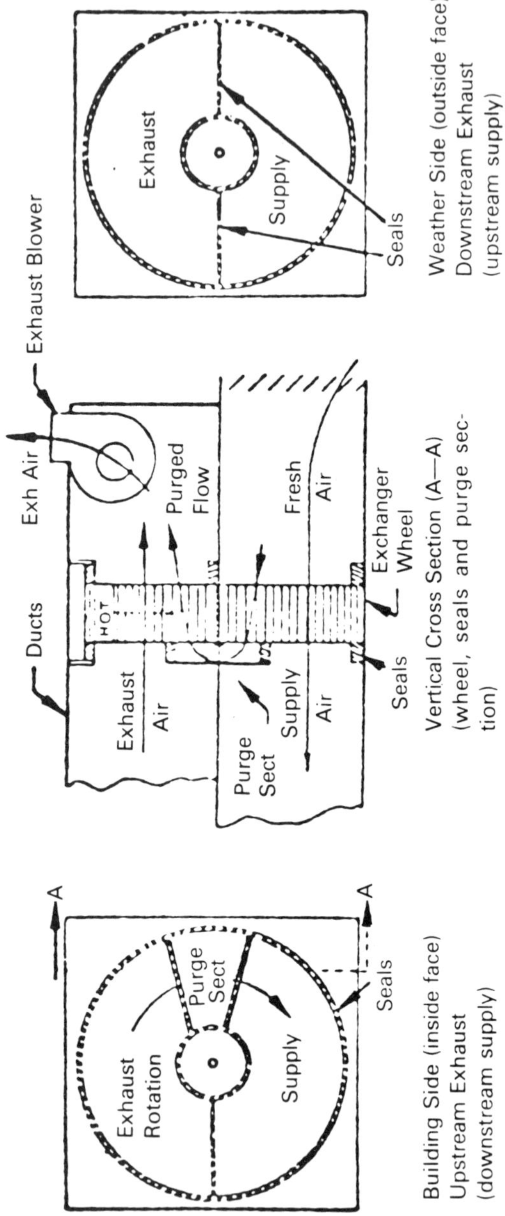

Source: PB-265 713

pipe operates on a nearly isothermal vapor cycle. Each heat pipe has within it a working fluid and a wick material. At the hot end energy is collected and vaporization of the working fluid occurs at the wick. This provides a pressure gradient that drives the vapor to the cool end. At the cool end, the vapor condenses releasing its latent heat of vaporization. The working fluid then returns to the hot end by diffusion through the wick. Figure 6.7 shows a typical heat pipe construction. Fins are usually added to the heat pipes in a heat pipe heat exchanger to enhance the heat transfer characteristics of the external surfaces. The individual heat pipes are then arranged in a tube bank to provide sufficient heat transfer.

FIGURE 6.7: HEAT PIPE

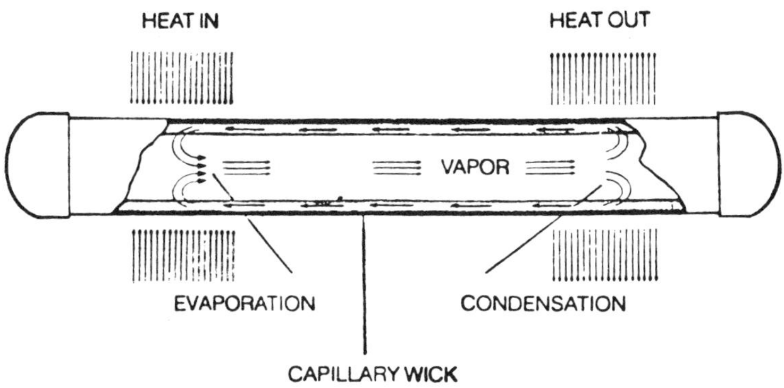

Source: PB-265 713

The heat pipe allows relatively large amounts of heat to be transported with small, lightweight structures. A significant advantage of this system is that no cross contamination occurs and no internal maintenance is required. This type of heat exchanger provides an excellent source of energy for hot process air requirements as well as combustion air preheat. Figure 6.8 shows a typical heat pipe heat exchanger.

Shell-and-Tube Heat Exchangers: Shell-and-tube heat exchangers are primarily liquid to liquid heat exchangers. Figure 6.9 shows a cross-flow once-through heat exchanger.

Shell-and-tube heat exchangers range from the simple example to the more complicated multipass units such as shown in Figure 6.10. Since no mixing of the fluids takes place, no contamination takes place. This type of heat exchanger is ideal for steam condensate and/or blowdown energy recovery. One fluid flows on the inside of tubes, while the

FIGURE 6.8: COUNTERFLOW HEAT EXCHANGER WITH HEAT PIPES

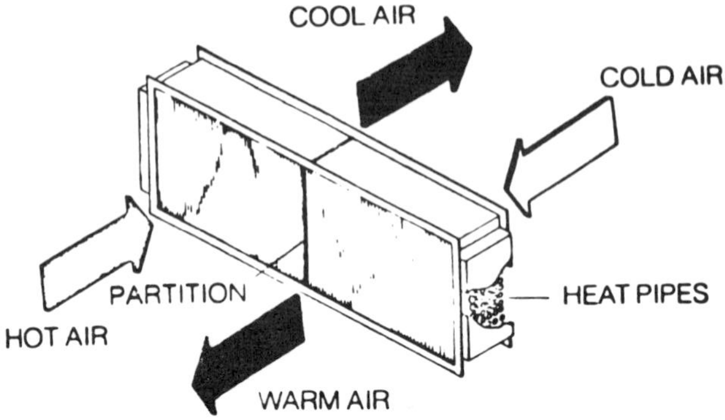

Source: PB-265 713

FIGURE 6.9: SINGLE SHELL-AND-TUBE HEAT EXCHANGER

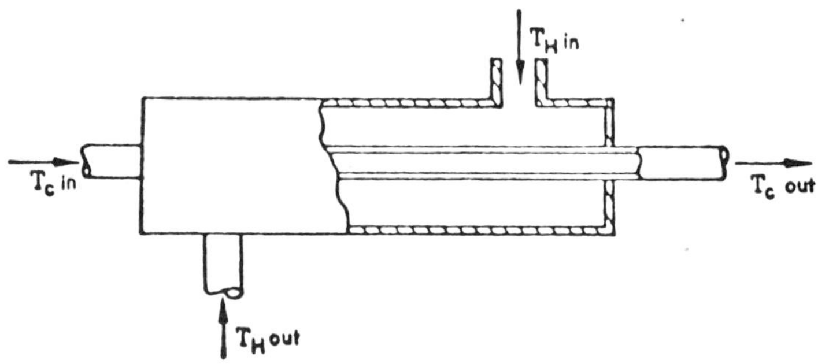

Source: PB-265 713

FIGURE 6.10: MULTIPASS SHELL-AND-TUBE HEAT EXCHANGER

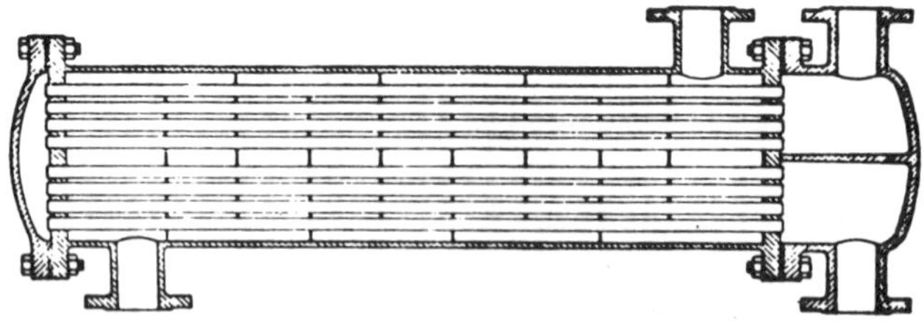

Two Tube Passes, One Shell Pass

Source: PB-265 713

other fluid is forced through the shell and over the outside of the tubes. To insure that the shell-side fluid will flow across the tubes and thus induce higher heat transfer, baffles are placed in the shell. Depending on the head arrangement at the ends of the exchanger, one or more tube passes may be utilized.

Conventional Finned Tube Heat Exchangers: The finned tube heat exchanger is perhaps the most common liquid to gas heat exchanger. In boilers with excessive flue gas temperatures a finned tube heat exchanger may be added to the boiler to preheat the feedwater and lower the exhaust gas temperature. Round tubes are connected together in bundles to contain the heated liquid and fins are welded or otherwise attached to the outside of the tubes to provide additional surface area for removing the waste heat in the gases. Figure 6.11 shows the usual arrangement for the finned-tube exchanger positioned in a duct and details of a typical finned-tube construction.

FIGURE 6.11: FINNED TUBE HEAT EXCHANGER

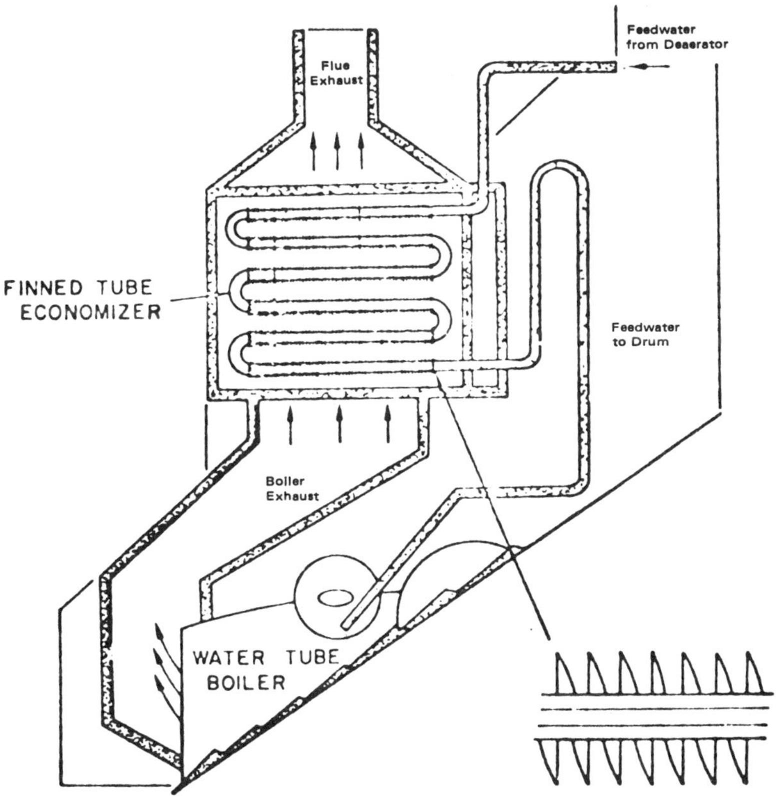

Source: PB-265 713

Energy Recovery from Excessive Steam Pressure

Throttling Turbines: In some instances it is impractible to generate steam at the pressure required by the process. A throttling valve is usually installed to drop the pressure to the required level. This pressure gradient can be utilized as a potential to do work through a throttling turbine.

Throttling or back pressure turbines can extract previously unused energy from the generated steam while meeting process requirements. Figures 6.12 and 6.13 show the arrangement of the throttling and condensing turbines. Figure 6.14 shows a comparison of the energy distribution in these two types of turbines.

FIGURE 6.12: BACK PRESSURE TURBINE

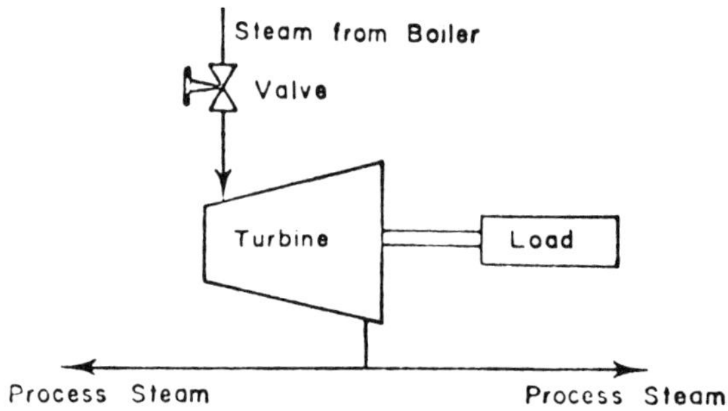

Source: PB-265 713

FIGURE 6.13: CONDENSING TURBINE

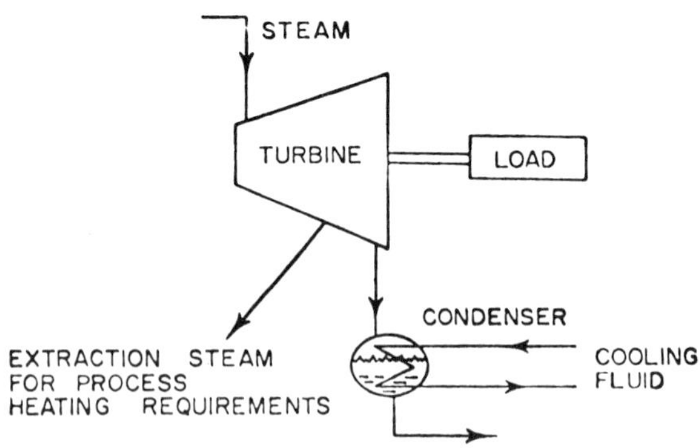

Source:PB-265 713

FIGURE 6.14: ENERGY DISTRIBUTION IN CONDENSING AND BACK PRESSURE TURBINES

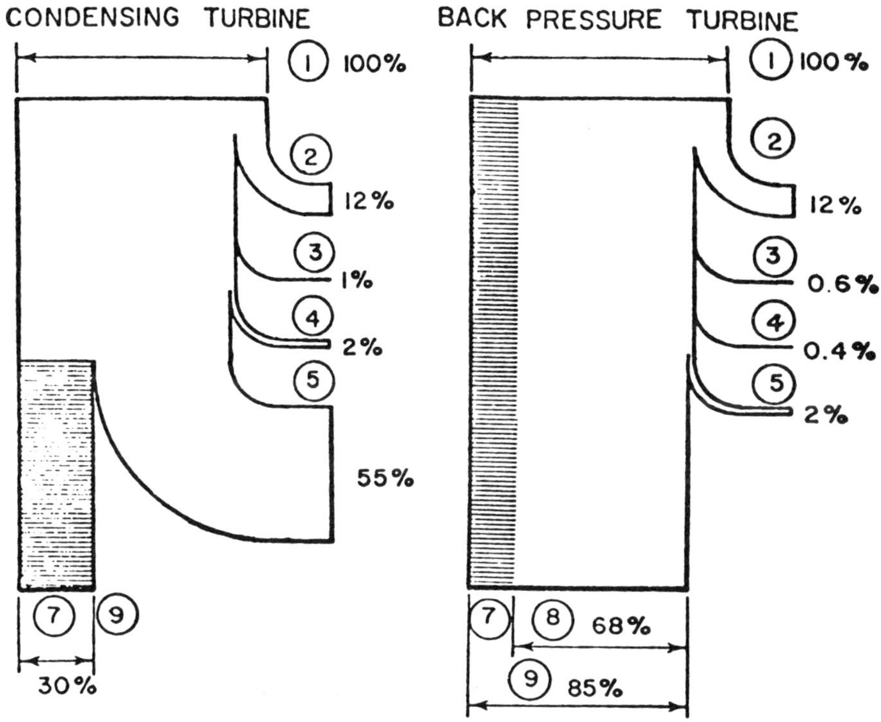

1. Thermal energy in fuel
2. Losses in boiler
3. Losses in generators
4. Losses in auxiliaries
5. Thermal energy lost to cooling water
6. Thermal energy lost through leakage conduction and radiation
7. Electrical energy produced by generators
8. Thermal energy from process steam
9. Available energy

Source: PB-265 713

Energy Recovery from Blowdown

As described in the first section of this chapter, the energy in the blowdown from boilers can be used to preheat feedwater. Two heat recovery systems for using the energy in the blowdown are discussed here.

Flash Tank System: The flash tank system shown in Figure 6.15

should be used when the expense and complexity must be reduced to a minimum. In this system, the blowdown is flashed by lowering the pressure in the flash tank. The steam produced is vented into the feedwater recovering approximately 50% of the energy released in blowdown.

FIGURE 6.15: BLOWDOWN SYSTEM WITH FLASH TANK

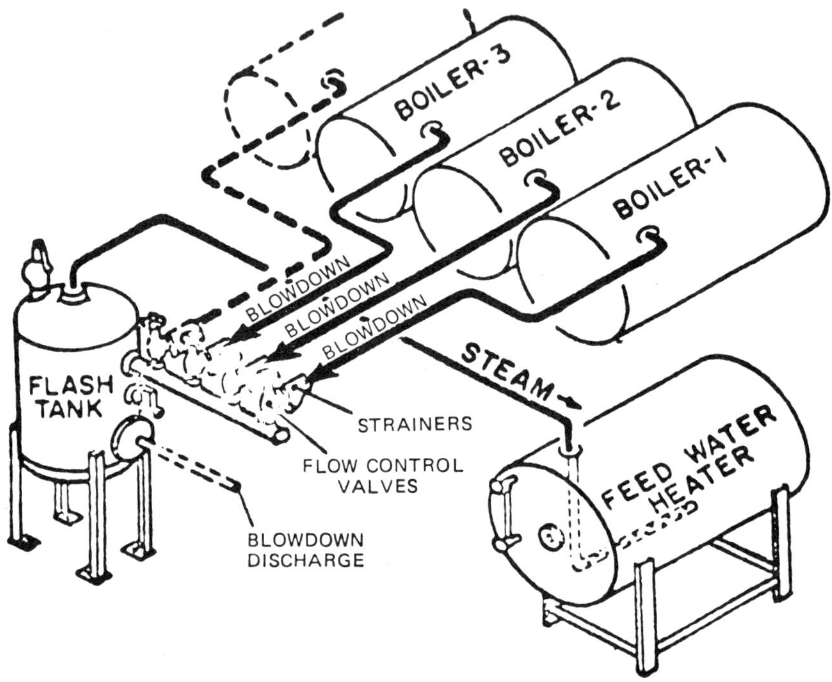

Source: PB-265 713

Flash Tank-Heat Exchanger System: This system consists of the flash tank and a heat exchanger as shown in Figure 6.16. The temperature of the blowdown leaving the flash tank is usually above 220°F. The additional feature in this system is the use of the energy in the liquid-solid mixture leaving the flash tank to heat make-up water as it flows through the heat exchanger. This results in an additional saving of approximately 25% of the energy in the blowdown.

The systems just described are summarized in Table 6.2.

FIGURE 6.16: BLOWDOWN SYSTEM WITH FLASH TANK AND HEAT EXCHANGER

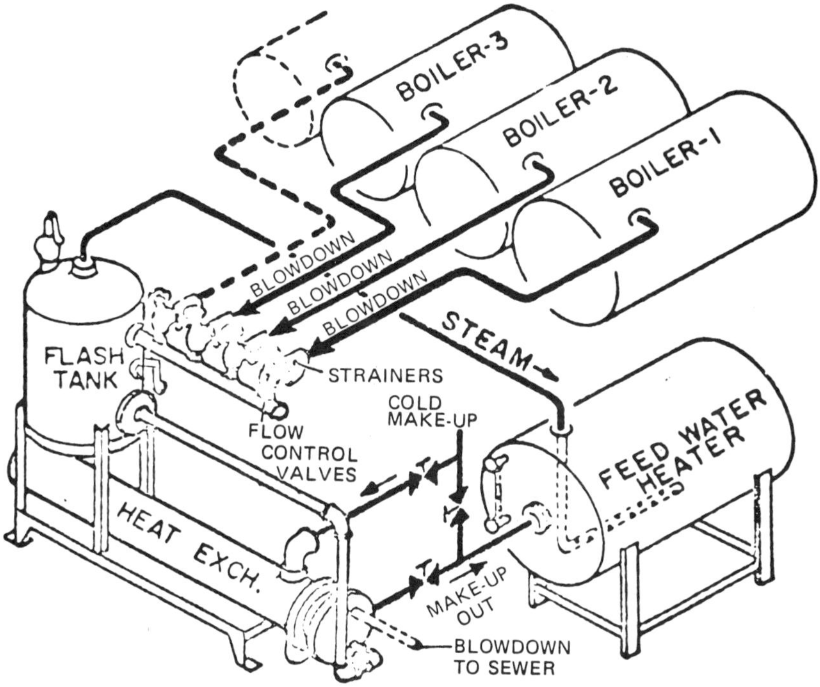

Source: PB-265 713

TABLE 6.2: ENERGY RECOVERY SYSTEMS

POTENTIAL ENERGY SOURCE	POTENTIAL ENERGY USE	AVAILABLE RECOVERY EQUIP.
Excessive Flue Gas Temperature	Air-Preheat	Heat-Wheel
	Heated Process Air	Heat-Wheel w/ Purge Section Heat Pipe Finned Tube
	Pre-heat Feedwater	Conventional Finned Tube
Steam Condensate and Blowdown	Pre-heat Feedwater	Shell-and-Tube
	Pre-heat Air Low Temperature	Conventional Finned Tube Heat Pipe
Excessive Steam Pressure	Use ΔP to produce work through turbine	Throttling Turbine

Source: PB-265 713

TECHNIQUES FOR ADJUSTING AIR/FUEL RATIO

Specific instructions for adjusting the A/F ratio are highly dependent on the type of burner control equipment. Some general guidelines regarding procedures for adjustments related to the measurement of specific fuel-product constituents are given here. The procedures are intended to describe safe adjustment techniques. Also, they will serve to reduce the time involved by outlining the most efficient steps to be taken.

The guidelines assume use of an Orsat analyzer, a heat prover, an oxygen analyzer, an infrared analyzer, or a gas-absorption unit to obtain near-stoichiometric combustion. Futhermore, near-stoichiometric combustion is defined as that A/F ratio which yields 0.5 to 1% O_2 in the flue products. An oxygen content of 0.5 to 1% oxygen was selected because 1) it is reasonably close to stoichiometric combustion so as to provide quite efficient combustion; 2) it is a value within the accuracy of the aforementioned instruments; and 3) it is on the lean side of stoichiometric combustion, thus ensuring the absence of dangerous flue products.

Using a Heat Prover

The A/F ratio is adjusted to obtain 0.5 to 1% oxygen in flue products. The heat prover is the most straightforward instrument to use in this situation since both percent oxygen and percent combustibles in the flue gases are measured. The first measurement obtained will define whether combustion is stoichiometric, i.e., both percent oxygen and percent combustibles indicate zero; rich, i.e., oxygen content is zero and combustibles are positive; or lean, i.e., percent oxygen is positive and percent combustibles is zero.

If combustion is stoichiometric or rich, the A/F ratio should be increased by reducing gas flow until 0.5 to 1% oxygen is obtained. If combustion is lean, reduce A/F ratio by incremental increases of fuel flow until 0.5 to 1% oxygen is obtained. Large changes to fuel flow should not be made until the changes in percent O_2 and in fuel flow are characterized by trials.

Note that, in going from the initial assumed condition of stoichiometric combustion to 0.5 to 1% oxygen, it is actually moving away from the most efficient condition. This is done solely for safety purposes. In short, one should not try to obtain stoichiometric combustion unless he is absolutely sure that ratio can be maintained over the entire operation range of the equipment.

Using an Oxygen Analyzer

The A/F ratio is adjusted to obtain 0.5 to 1% oxygen in flue products. Since only O_2 is being measured, if the meter should indicate no oxygen present in the flue products combustion is either rich or stoichiometric. In either case the gas flow should be reduced until oxygen is detected. Then the gas flow should be adjusted carefully to establish 0.5 to 1% oxygen in the flue products.

If greater than 1% oxygen is measured, the gas flow should be increased incrementally to reduce oxygen to within 0.5 to 1%.

Using Infrared CO_2 and CO Analyzers

The A/F ratio is adjusted to the corresponding CO_2 level to give 0.5 to 1% oxygen in flue products, i.e., 11.6 to 11.1% for natural gas.

When the only measuring device being used is the infrared CO_2 analyzer, information should be available regarding the theoretical flue-gas analysis of the fuel being used. The fuel-product analysis of methane is used in the following example.

With a CO_2 measurement of 10%, combustion could be rich at a ratio of approximately 8.7. Combustion could also be lean at a ratio of about 10.9.

The first adjustment should be made toward lean combustion by reducing gas. If CO_2 decreases, then combustion is definitely lean. If CO_2 increases with a gas decrease, then combustion is rich. Gas decreases should be small to avoid large changes in CO_2 and the possibility of passing through stoichiometric conditions without detection. Under rich combustion conditions, the gas should be decreased until CO_2 content reaches stoichiometric conditions at 11.8%. Then the gas should be reduced further until CO_2 drops to about 11.4%, which is about equivalent to 1% oxygen. With initial lean combustion conditions, the gas should be increased until CO_2 rises to 11.4%.

If flue-product analysis information for the particular fuel being used is not available, the above procedures will suffice, but careful attention must be paid to detect how gas-flow changes affect CO_2 content. All fuels will have peak CO_2 content at stoichiometric conditions. Once peak CO_2 content has been detected, the final CO_2 content should be adjusted to about 5% less than peak value by gas-flow reduction. This will ensure operation on the lean side of stoichiometric combustion.

The downfalls of using a CO infrared detector are obvious. If CO is detected, gas flow should be reduced until no CO is measured. It is not possible, however, to determine how much gas reduction is necessary to provide 0.5 to 1% oxygen since no CO occurs above stoichiometric ratio. Relying solely on a CO measurement should therefore be avoided.

Using an Orsat

The A/F ratio is adjusted to obtain 0.5 to 1% oxygen in flue products. Although all products of combustion can be measured with an Orsat, only CO_2 and O_2 need be measured to accomplish accurate adjustment.

Measuring both O_2 and CO_2 will clearly define the A/F ratio if the theoretical flue-gas analysis is used. If no oxygen is detected, the actual A/F ratio is determined from the CO_2 content on the rich combustion side of stoichiometric ratio. Then the gas should be reduced until 0.5 to 1% oxygen is obtaned. If oxygen is detected in the first measurement, then the gas is simply adjusted to obtain between 0.5 and 1% oxygen.

Since sampling with an Orsat is noncontinuous, sufficient time must be allowed between adjustments to gas flow and subsequent sampling. This will ensure samples representative of actual conditions after the A/F ratio change.

Using an Absorption Device

Either CO_2 or O_2 is measured to obtain 0.5 to 1% oxygen or its equivalent in fuel products. The procedures to be used with these devices are similar to those discussed for the oxygen analyzer or infrared CO_2 analyzer. The precautions on sampling discussed in the Orsat section should also be observed.

Using Air and Fuel Meters

Combustion ratios are adjusted to obtain approximately 0.5 to 1.0% oxygen in flue products. Using air and fuel meters to set up combustion is not the most reliable method. The effects of leakage and burner performance cannot be accounted for. The following procedures are suggested if meters are supplied, meter calibration curves are available, and no flue-product-analysis equipment is available.

If using meter data only, one must have flue-product-analysis information for the particular fuel being used and fuel and air orifice flow charts. If natural gas is the fuel, the stoichiometric A/F ratio is defined as 9.4 to 1. Also, 0.75% oxygen is obtained at an A/F ratio of approximately 10.4 to 1. This ratio should be established in the actual combustion process by first measuring the air flow at high-fire conditions.

The required gas flow is then air flow divided by 10.4. The required gas-pressure drop across the meter can be determined by referring to a gas flow chart. The gas flow is then adjusted to give the correct pressure difference. At this point, flue-gas analysis should be checked with an instrument to ensure that the actual combustion is representative of measured inputs.

REFERENCES

(1) Betz, W.H. *Betz Handbook of Industrial Water Conditioning.* Third Edition, Philadelphia, 1950.
(2) Salisbury, J.K., *Kent's Mechanical Engineers' Handbook,* John Wiley and Sons, Inc., 1967.

Efficiency Improvement Equipment

The material in this chapter was excerpted from a report by KVB, Inc. (PB-262 577).

This chapter discusses the auxiliary equipment and boiler components currently available to improve boiler efficiency. Most of this equipment can be retrofitted to existing units but in many cases it is economically more practical to purchase it with a new unit.

AIR PREHEATERS

Principle of Operation

Air preheaters are employed to transfer heat energy from the stack gas to the incoming combustion air supply. Boiler efficiency is improved at all loads as a result of reduced stack gas temperatures. Improved combustion conditions at the burner permitting operation with lower excess air can be an additional benefit.

Air preheaters can be used economically on units with outputs as low as 25,000 pounds of steam per hour (1). The choice between an air preheater and an economizer for larger units is made on the basis of

- stack gas temperature
- installation and operating costs

- draft losses
- type and arrangement of the boiler
- operating steam pressure that may limit the maximum feedwater temperature and thereby the application of an economizer
- flue gas corrosive properties (i.e., fuel type) which may determine the lowest practical stack temperature
- maintenance requirements.

A dual installation incorporating both an air preheater and economizer is often used on larger units with steam pressures above 400 psig.

Performance

Considering only the benefit of utilizing heat energy in the flue gas to heat the combustion air, the overall steam generator efficiency will increase approximately 2.5% for every 100°F decrease in exit gas temperature or 2% for every 100°F increase in combustion air temperature. Any accompanying improvements in burner excess air with preheated combustion air will enable additional efficiency gains (see Figure 7.9, p. 138). For example, an air preheater providing combustion air of 300°F can result in a fuel savings of up to 5%.

The efficiency improvement for a given drop in flue gas temperature can be estimated using Figure 7.1. Using this value, an approximate gross annual fuel savings can then be estimated using the nomograph shown in Figure 7.2.

FIGURE 7.1: APPROXIMATE EFFICIENCY IMPROVEMENT WITH DECREASED FLUE GAS TEMPERATURES (Ref. 2)

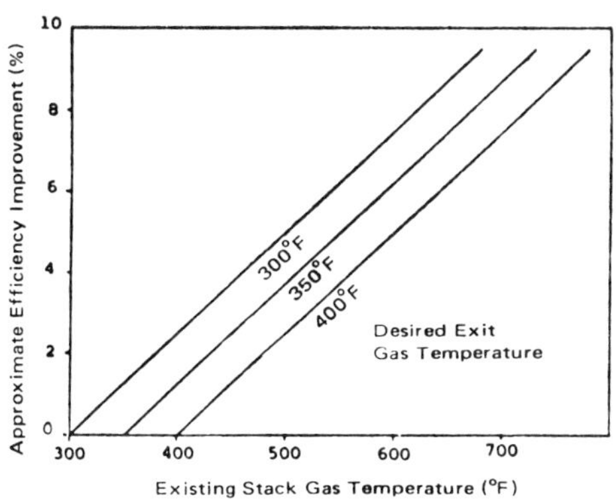

Source: PB-262 577

FIGURE 7.2: POTENTIAL GROSS SAVINGS IN ANNUAL FUEL COSTS FOR A GIVEN EFFICIENCY IMPROVEMENT*

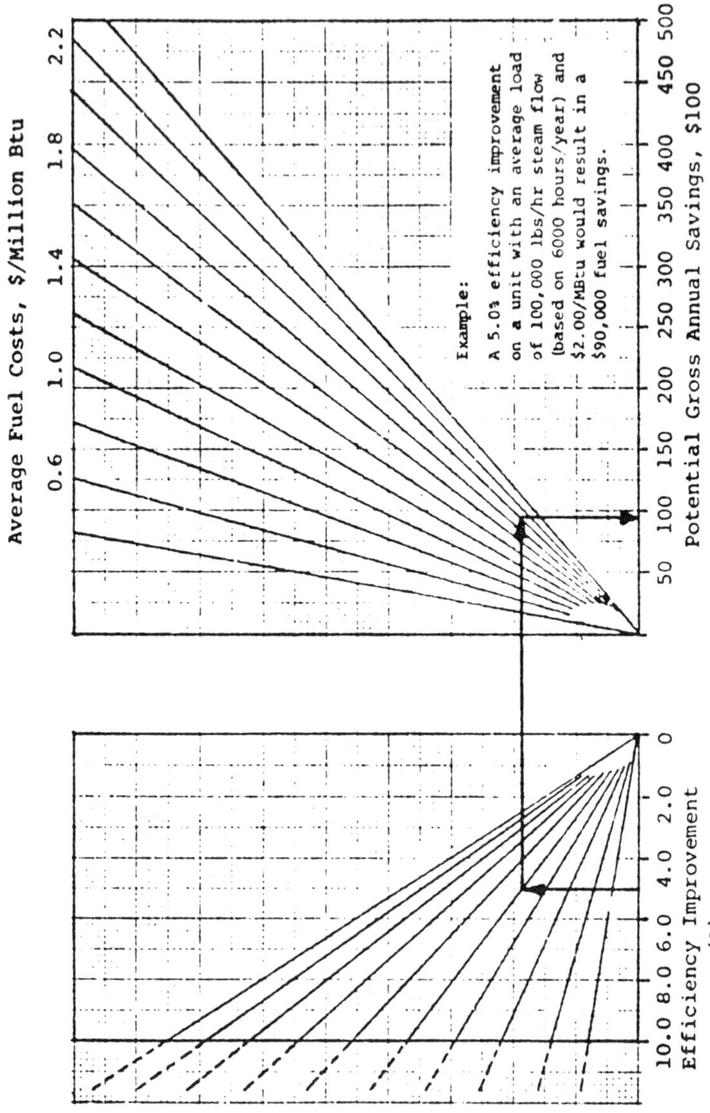

*(Based on 6,000 hr/yr Operation)

Source: PB-262 577

The steam flow used in this nomograph should correspond to an average load based on 6,000 hours/year operation. Subtracting fixed charges and increased operating costs will result in an approximate net annual savings.

Costs

The total cost of installing an air preheater will of course vary from boiler to boiler. The approximate range of capital equipment and installed costs are given in Figure 7.3. The capital costs will vary depending on the pressure drop that can be tolerated, air heater design, materials and desired performance. Installed costs can be as much as 3 to 5 times the capital costs and are affected by

- available space
- required equipment repositioning
- duct modifications
- air fan changes if necessary to overcome increased pressure drops
- piping and wiring changes
- additional sootblower requirements.

FIGURE 7.3: APPROXIMATE AIR PREHEATER EQUIPMENT AND INSTALLED COSTS

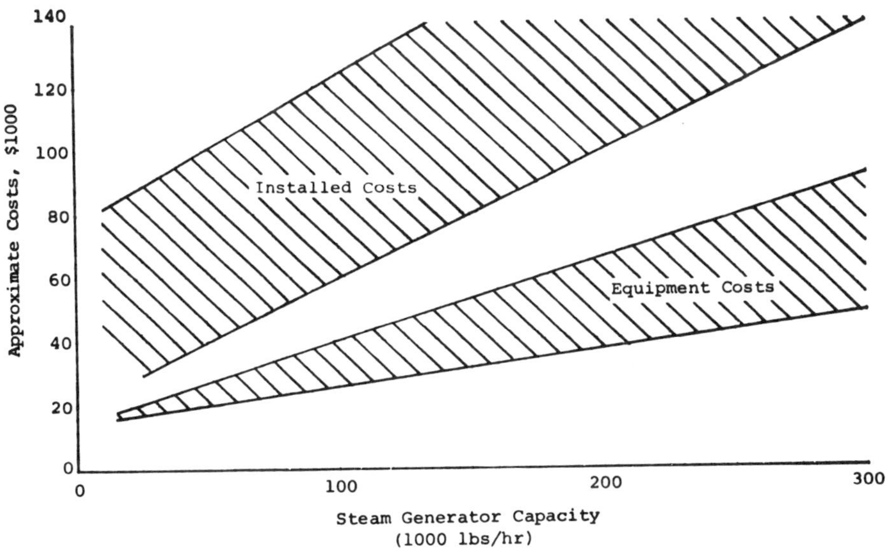

Source: PB-262 577

Significant annual costs must also be considered including fixed costs, operating and required maintenance costs. Fixed and operating costs may amount to 10-20% of the installed costs. A cold end heating surface replacement can cost another 10% of the equipment costs.

Types of Air Heaters

Air preheaters are classified as either recuperative or regenerative. In the recuperative type, the heat energy is transferred directly from the flue gas on one side of a stationary separating surface to air on the other side. This separating surface may be either a tubular or flat plate design. Typical arrangements for the more common tube types are shown in Figure 7.4. Also included in this class are recuperative steam coil heaters that are sometimes used to preheat the combustion air before entering the main air preheater. This is occasionally necessary to reduce "cold side" corrosion by increasing the minimum tube metal temperatures exposed to the flue gas. Advantages of recuperative air preheaters include

- adaptable to a wide variety of physical arrangements
- limited cross contamination between flow streams
- no external power
- no moving parts.

A disadvantage is the adverse effects of soot deposits on heat transfer efficiency. Use of the plate type is diminishing due to cleaning difficulties.

The regenerative type preheater transfers the heat energy from the flue gas to the combustion air through an intermediate heat storage medium. Steam generating units most commonly use a rotary "wheel" design shown in Figure 7.5. Advantages of this type of unit include

- adaptable to several duct arrangements
- compact and lighter than comparable recuperative units
- short flow passages reducing fouling
- heat transfer characteristics not affected by soot deposits
- can operate at a lower exit gas temperature than recuperative units

FIGURE 7.4: TYPICAL RECUPERATIVE AIR PREHEATER ARRANGEMENTS (Ref. 4)

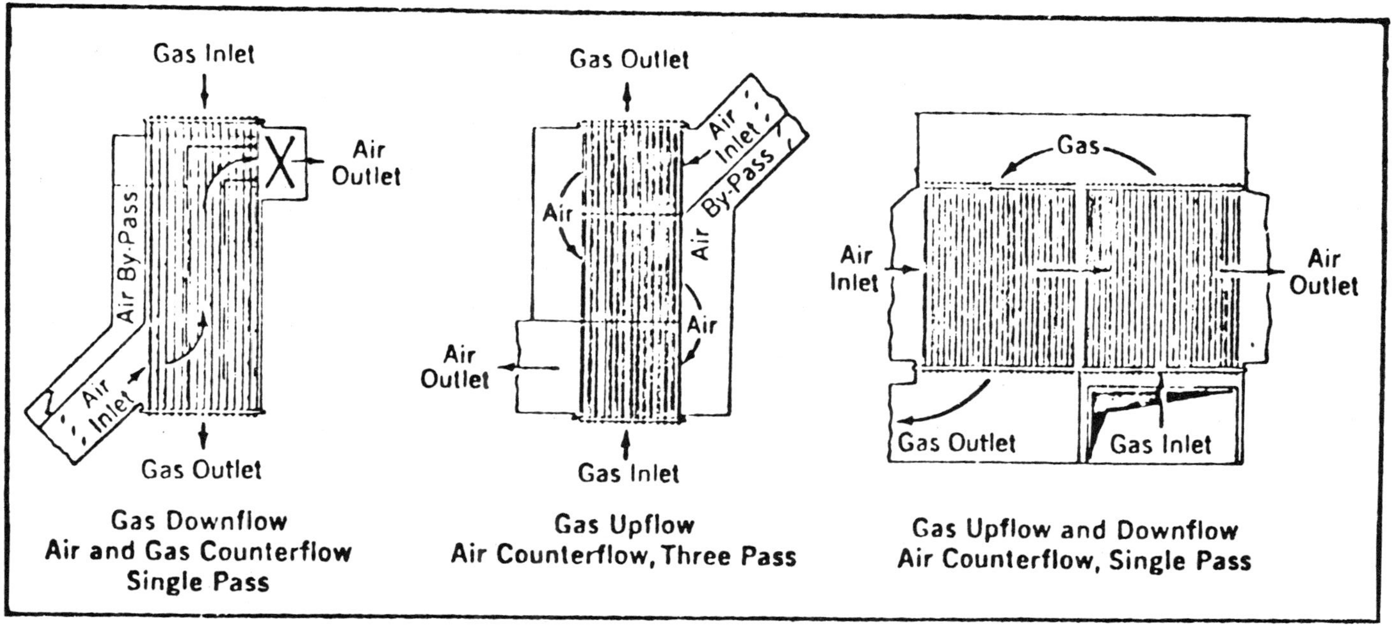

Gas Downflow
Air and Gas Counterflow
Single Pass

Gas Upflow
Air Counterflow, Three Pass

Gas Upflow and Downflow
Air Counterflow, Single Pass

Source: PB-262 577

The disadvantages are

- cross contamination around seals
- external power required

It is generally more common to use a regenerative rotary unit on retrofit installations.

FIGURE 7.5: REGENERATIVE AIR PREHEATER (Ref. 4)

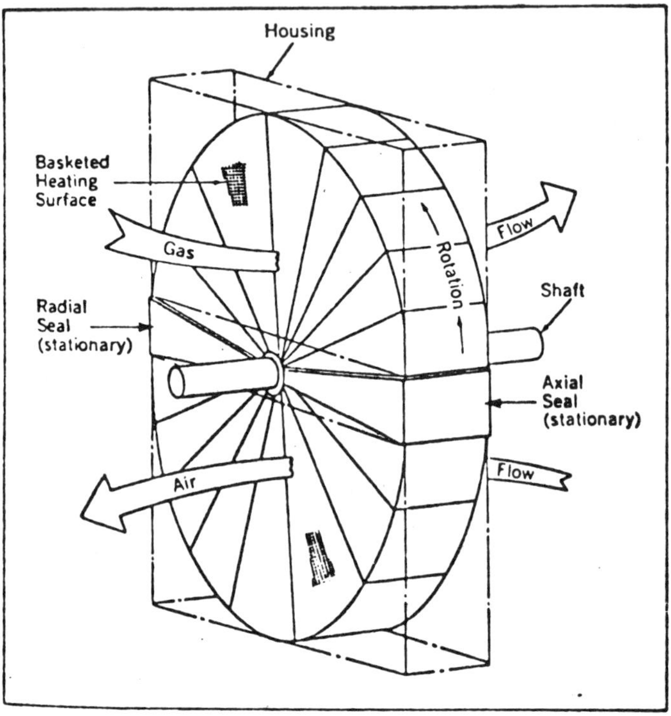

Source: PB-262 577

A third type of air preheater not currently in use on steam generating units is a heat exchanger employing heat pipes as shown in Figure 7.6. Heat pipes are a relatively new technological development and offer a promising alternative to conventional heat transfer devices. This type of system may be very adaptable to industrial boilers and offers the following advantages:

- no cross contamination
- no external power or moving parts
- minimum maintenance

The primary disadvantage is a 600°F maximum operating temperature which would limit its application. Flue gas bypasses are recommended for protection from possible corrosion at low loads and overheating of the heat pipe element at high loads.

FIGURE 7.6: HEAT PIPE HEAT EXCHANGER (Ref. 5)

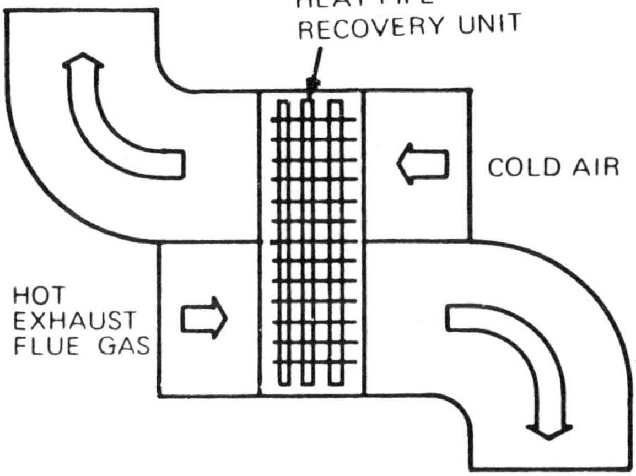

Source: PB-262 577

Application Considerations

Impact on Emissions: Increased combustion air temperature is known to have an adverse effect on emissions of oxides of nitrogen (NO_X), particularly for natural gas fuel. For a range of industrial boiler designs, the NO_X increased by 20 ppm to 100 ppm per 100°F increase in combustion air temperature (6). This factor must be considered where existing or proposed air pollution regulations may be applicable. Improved combustion conditions in the furnace resulting in lower excess air requirements may tend to offset somewhat this increase in NO_X levels since NO_X usually decreases with lower excess air.

Use of preheated combustion air also appears to affect other air pollutants. Lower SO_3 emissions have been reported; however this has not been fully substantiated (1). More complete combustion in the furnace also reduces particulate and CO emission levels.

Corrosion: The lowest acceptable flue gas temperature at the exit of the air preheater is limited to prevent moisture condensation on heat transfer surfaces which would lead to sulfuric acid corrosion and expensive repairs. Factors affecting these minimum temperature limits include

- sulfur content of the fuel
- moisture content of the flue gas
- type of fuel
- ambient air temperature

Recommended minimum average cold end temperatures developed by C-E Air Preheater are presented in Figure 7.7 (7). Note that the minimum metal surface temperatures are determined by averaging the exit gas and entering air temperature of 80°F, the exit gas temperatures corresponding to 235°F and 175°F average cold end temperature would be 390°F and 270°F respectively.

Average cold end temperature may be controlled by one or more of the following:

- Bypass a portion of inlet air around heating surfaces
- Recirculate a portion of the hot air from the air heater outlet into the air entering the heater
- Use of a recuperative steam coil in the air duct upstream of the main air preheater.

Corrosion may be further reduced by using acid-resistant metals such as cast iron and corrosion-resistant steel or by coating metal surfaces with enamel.

Ash deposits on the heat transfer surfaces also contribute to corrosion. The porous ash reduces underlying metal temperatures and allows the flue gas moisture to condense on the metal surface. The moisture that results combines with the acid in the ash to promote rapid corrosion. Soot blowers employing steam or air as cleaning agents are used to remove the ash deposits.

The life of the heat transfer surfaces in an air preheater is very dependent on the prevention of cold end corrosion and the sulfur content of the fuels used. The expected life ranges from 4 to 5 years with high sulfur fuels to 20 years on natural gas.

FIGURE 7.7: COLD END TEMPERATURE AND MATERIAL SELECTION GUIDE

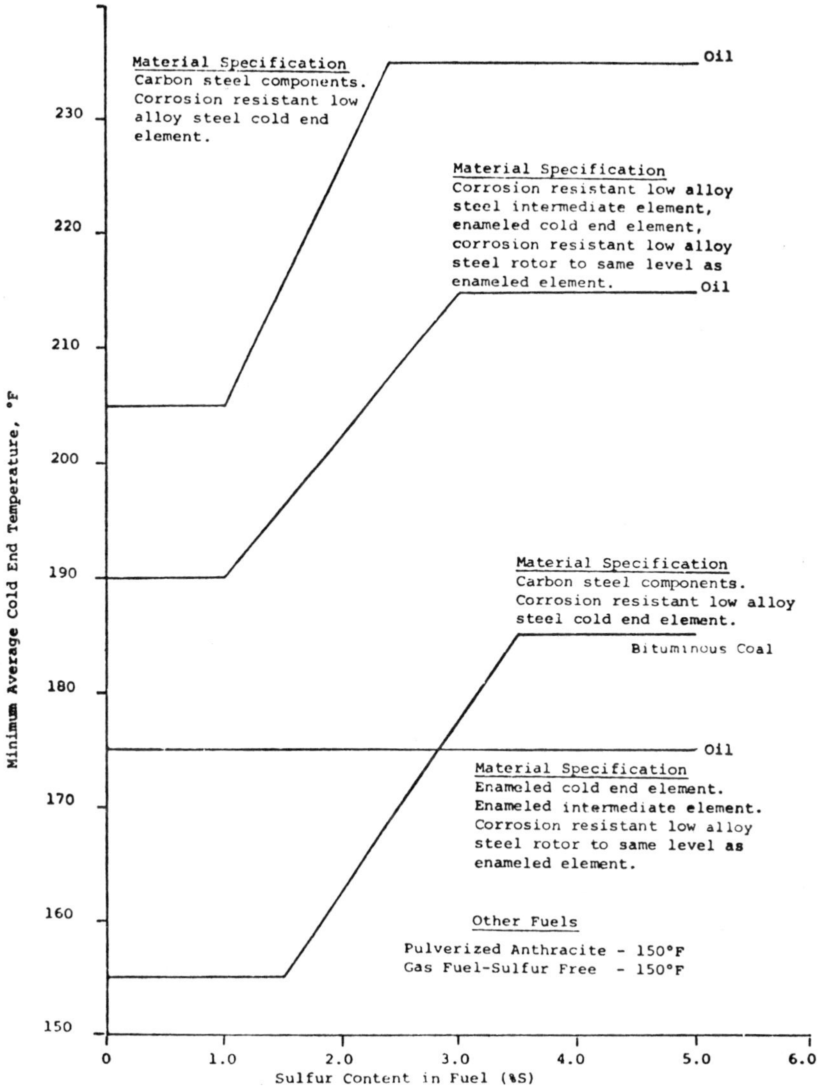

Source: PB-262 577

Maximum Combustion Air Temperatures: The temperature to which the combustion air can be preheated is limited by the entering gas temperature, the firing equipment and the type of duct, windbox and furnace construction. Maximum combustion air temperature is generally taken to be 250°-350°F for stoker-fired units. Pulverized coal firing may require preheated air up to 700°F (1)(2). Retrofit applications must consider these limitations along with required insulation of combustion air ducts.

Air Preheater Size: The physical dimensions of the air preheater are determined by arrangement and dimensions of air and gas passages, space available for the air preheater, type of fuel fired, desired final combustion air and flue gas temperature, air and gas flow requirements, and pressure drop requirements.

Forced Draft and Induced Draft Fan Modifications: An evaluation of the use of air preheaters in existing units must consider the added cost of revamping forced and induced draft fans to accommodate the added resistance to air and gas flow. Static pressures will also increase due to changes in air and gas temperatures throughout the unit. Additional fan power will be required to make up for leakage around seals on a rotary type unit. Typical design leakage can be as much as 10-15% of the exit gas flow on this type of air heater while leakage on recuperative heaters can be eliminated entirely by replacing or repairing leaking elements.

A complete engineering evaluation will be required to determine the feasibility of using existing or required new equipment and the added operational costs.

Additional Desirable and Undesirable Aspects: In addition to the improvements in overall boiler efficiency, and subsequent fuel savings, air preheaters offer the following desirable features:

- Improve combustion efficiency resulting in reduced excess air requirements for complete combustion and total gas flow through the unit.
- Aid in stabilizing fuel ignition which permits greater load flexibility including improved low load operation.
- Increase furnace temperature and heat absorption rate resulting in greater steam producing capacity limited by drum steam separating capacity.
- Result in more complete burning of the fuel producing cleaner gases passing through the unit, less fouling and fewer boiler outages for cleaning.

The disadvantages of air preheaters include the following:

- Increased stoker and furnace refractory maintenance costs.
- Accumulation of deposits in the gas passages of the air heater may restrict gas flow and in extreme cases may ignite causing serious damage (if poorly maintained).

ECONOMIZERS

Principle of Operation

An economizer is an arrangement of feedwater tubes located in the exhaust duct that absorb a portion of the heat energy that would otherwise be lost in the flue gas. This recovered energy provides additional heating of the feedwater thereby reducing the boiler firing rate necessary to generate steam, improving overall boiler efficiency. Economizers can also be employed to heat hot water for space heating or process requirements apart from the boiler water cycle.

Economizers are generally preferred over air preheaters in retrofit additions to industrial sized units for the following reasons:

- lower initial capital costs
- no impact on NO_X emissions
- lower draft losses
- minimal auxiliary power requirements (slight increases in feedwater pumping and combustion air fan requirements).

An economizer is more economical than an air preheater for small, low pressure boilers with outputs below 50,000 pounds of steam per hour (1). Air preheaters become competitive with economizers for larger units and the choice is made on the basis of the factors as itemized in the Air Preheaters Section.

Performance

An increase in feedwater temperature of 10°F will result in a fuel savings of approximately 1.0% (2). The practical extent to which the feedwater temperature can be increased using an economizer is limited by several factors including

- flue gas temperature
- steam pressure

- economizer surface area
- corrosion characteristics of flue gas.

The efficiency improvement with an economizer is reflected in lower stack gas temperatures and can be estimated using the graph presented in Figure 7.1. The relationship between stack temperature reduction and feedwater temperature increase can be closely estimated on the basis that a three degree decrease in fuel gas temperature will raise the feedwater temperature by one degree. There is a maximum feedwater temperature depending on the operating pressures of the boiler.

The gross annual fuel savings can be estimated from the nomograph presented in Figure 7.2 using the projected efficiency improvement, average steam flow and fuel costs. Subtracting fixed charges and increased operating costs will result in an estimated net annual savings.

Costs

The approximate equipment and installed costs of retrofitting an economizer into an existing steam generator are given in Figure 7.8. These costs will vary for each application. Capital costs are influenced by such factors as

- arrangement of economizer tubes
- desired performance
- tube size and materials.

FIGURE 7.8: APPROXIMATE ECONOMIZER EQUIPMENT AND INSTALLED COSTS

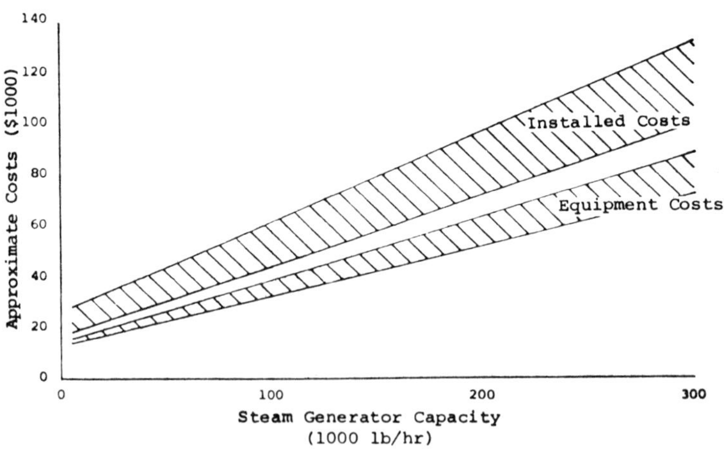

Source: PB-262 577

Installed costs will typically vary from 30 to 100% of the capital costs depending on:

- available space
- duct modifications
- required feedwater preheaters
- plumbing and pumping alterations
- air fan changes due to increased draft loss

Maintenance costs are generally limited to internal and external cleaning. Tube replacement costs can be minimized by proper selection of economizer tube materials for given flue gas conditions.

Design Factors

This section discusses basic economizer design factors and other items to be considered in employing an economizer to improve unit performance.

Size: Heat transfer rates are highest when the temperature differential between the flue gas and the feedwater are greatest. Since the feedwater is at a lower temperature than the water in the convective pass tubes of the boiler, absorption of heat from the flue gas is potentially more effective in the economizer than in the convective surfaces of the boiler. Current boiler design trends have made use of this fact by increasing economizer surface with corresponding decreases in boiler surface (1). The economizer surface area is generally 25 to 30% and 40 to 60% of the boiler heating surface for applications with and without air preheaters respectively. Economizer surface area is limited by increased draft losses and outlet temperature differential between the gas and feedwater.

Arrangement: Economizers can be classified as either integral or separate. Integral economizers are arranged in vertical or horizontal banks of tubes located within the boiler casing. Vertical tubes are similar to a bank of boiler tubes and are frequently connected at both ends with drums. Horizontal tubes are arranged in closely spaced, staggered rows. Spacing is limited to a minimum of 1¾ to 2 inches to prevent ash accumulation from plugging the economizer and to maintain draft losses to less than 4 inches of water. Horizontal tubes have a continuous loop arrangement.

Separate economizers are located downstream of any convective region and outside the boiler casing. They are generally arranged horizontally. Retrofit economizers usually employ separate economizers. Water flow in the economizer is normally from the bottom upward through the tubes allowing any steam that may be generated to be bled off the top. The flue gas passes either down along the tubes moving counter to the water flow or across the tubes in single or multiple passes. The counterflow arrangement is preferred as it reduces required surface area and results in lower draft losses. Baffles and bypass ducts are sometimes employed to control gas and feedwater temperatures.

Tubes: Economizer tubes are generally made of cast iron or steel and range from 1¾ to 2 inches o.d. Steel tubing is employed in high pressure applications and offers the advantages of thinner tubing, closer spacing, higher conductivity, and lower initial cost.

Cast iron is used in low pressure applications especially where corrosion is expected. A combined steel and cast iron design is sometimes used where maximum resistance to both internal and external corrosion is desired.

Extended tube surfaces are often used to increase heat absorption. The type of extensions and spacing are determined by the fuel characteristics and provision for cleaning. Short, widely spaced fins are employed with coal and residual oil while tall, closely spaced fins are used with natural gas or other clean fuels. The advantages of this type over bare surfaces include lower initial costs, strength against bending, and reduced tubing requirements.

Extended surface tubes are normally installed in horizontal banks in a staggered arrangement.

Economizer Outlet Temperature: The maximum economizer outlet temperature is usually maintained 35 to 75 degrees below the saturation temperature to prevent steaming, water hammer and thermal shock. In certain applications, steaming economizers are employed to produce part of the steam. They are not normally employed in industrial boilers where a high feedwater make-up is necessary.

In the case of retrofit economizers significant feedwater temperature alterations can have a detrimental effect on boiler water circulation. Industrial boilers generally use "natural" water circulation (i.e., circulation throughout the boiler is maintained by variations in water densities caused by heat absorption in various parts of the boiler circuit). Changes in feedwater temperature can result in a lower circulating head and some circuits may not maintain sufficient flow for proper cooling. This may require modifications to existing boiler heat transfer passes.

Internal Corrosion: Economizers are subject to internal corrosion from dissolved oxygen and low hydroxyl ion concentration. As the temperature of the feedwater increases, the oxygen is driven out of solution and attacks the tube interior surfaces. The feedwater is therefore preheated to 220°F by the use of deaerator heaters to remove dissolved oxygen. A pH value of between 8 and 9 is normally maintained in the feedwater by recirculating boiler water into the inlet of the economizer.

External Corrosion: The same external corrosion considerations that were discussed for air preheaters apply to the application of an economizer. Since the tube metal temperature is very close to the feedwater temperature, the inlet feedwater temperature is normally maintained at 220°F or higher so that the tube metal surfaces will be above the minimum cold end temperatures presented in Figure 7.7. Exit flue gas temperatures are at least 80°F above this to provide sufficient heat transfer to the feedwater.

Air Pollution Emissions: Air pollution emissions are generally not affected by the use of an economizer. CO, particulate and combustible emissions may be reduced in applications where overfiring of the boiler was necessary to maintain peak load conditions prior to incorporation of an economizer.

FIRETUBE TURBULATORS

Principle of Operation

Firetube turbulators are small baffles inserted into individual firetubes in the upper passes of firetube boilers to induce turbulence in the hot gas stream and thereby increase the convective heat transfer to the tube surface. These devices have also reportedly enabled installers to balance gas flow through the firetubes to achieve more effective utilization of existing heat transfer surface. The corresponding efficiency improvement at a given boiler firing rate and excess air level are reflected in lower stack gas temperature and increased steam generation.

Performance

As described previously, a reduction of 100°F in stack gas temperature will result in a 2½% efficiency improvement. Manufacturer claims range up to 10%, which would correspond to a 400°F reduction in flue gas temperature. These very high improvements are probably a combined effect of reducing excess air levels, lower stack gas temperatures and tube cleaning which may be performed during installation of the turbulators.

Costs

Installed costs will vary with allowable pressure drops, boiler capacity, and configuration. These costs will range to $500 for units below 16,000 pounds per hour and up to $2,000 for larger units.

Physical Description and Application

Turbulators may consist of angular metal strips inserted into the second or second and third flue gas passes or spiral metal blades and baffles installed at the tube inlet. Those of the first type are generally two to eight feet in length and sized to fit the inside diameter of the tubes.

These devices have been generally employed on firetube units with steam capacities less than 30,000 pounds per hour as substitutes for more costly economizers or air preheaters. Turbulators can be employed in gas and oil fired boilers with two and three pass systems but are not recommended for coal fired units.

Turbulators increase convective heat transfer to the tube surface by increasing flue gas turbulence. By selectively altering the flow resistance through individual tubes, they can also be employed to balance gas flows through the tubes.

Allowances must be made to overcome the increased pressure drop through the system and assure sufficient air flow for complete combustion. Often this will require readjustments of burner controls following installation of the turbulators. The need for these adjustments will be determined by a post-installation efficiency check (O_2, CO, stack temperature measurements) which should be performed to verify complete combustion.

Air Pollution Emissions: Manufacturers claim that the improved combustion conditions will result in reduced combustibles in the flue gas.

COMBUSTION CONTROL SYSTEMS AND INSTRUMENTATION

Principle of Operation

Combustion controls regulate the quantity of fuel and air flows in a steam generating unit. The primary objectives of these controls are to

- Provide adequate heat input to meet steam demands automatically
- Protect personnel and equipment
- Minimize pollution, and
- Minimize fuel usage

The first of these objectives is met by sufficient fuel input while the latter three are dependent on maintaining the proper air flow with respect to it.

An optimum range of excess air exists for each combination of fuel, firing mode and furnace condition. Too little air flow results in combustible emissions which are hazardous, polluting and can contain considerable amounts of unused energy. The minimum safe excess air level for conventional burners is generally taken to be 10% for gas fuel and 15% for oil fuel; however specific applications may require higher levels. Too much excess air results in higher NO_x emission levels and increased mass flow and energy losses up the stack. Heat transfer within the unit is also reduced with high excess air levels culminating in higher stack gas temperatures and further performance losses.

All boilers have some type of combustion control arrangement from simple manual control to a highly sophisticated computerized set up. The choice of the type of control system is made on the basis of

- steam generating unit capacity and turndown required
- steam demands and expected fluctuations in steam flow
- expected performance levels that will require more sophisticated systems for higher operating efficiency
- costs
- pollution regulations that may require low excess air operation to minimize NO_x emissions
- adequate safety interlocks.

It has been shown to be economical to equip units as small as 25,000 pounds of steam per hour with fuel saving control systems (8). Caution, however, must be exercised to avoid selecting a control specification that may be operationally impractical.

Instrumentation, in contrast to combustion controls, consists of passive elements that present the boiler conditions as they exist but are unable to detect deviations from desired operation and make corrective action without control systems or operator involvement. Without such further information, it is impossible to determine whether the equipment is operating to the best advantage or if corrective action is necessary to restore performance.

Use of a continuous oxygen analyzer to determine excess air levels is highly desirable with regards to pollution control and energy conservation. It may be used either to provide information to the operator on control system operation or as an integral part of the combustion control system.

Performance

The increase in boiler efficiency using a retrofit combustion control system will be dependent on the type and condition of the existing equipment. Table 7.1 (8) presents the sources of errors and the expected excess air operating levels as per manufacturers for each of the six control systems that will be described. With the simplified positioning control (jackshaft) system used as the basis for comparison, installation of a sophisticated control system that adjusts air flow in relation to fuel flow with excess O_2 correction is shown to reduce air levels on gas fuel from 43 to 10% excess air. The associated improvement in efficiency can be determined for gas, oil and coal fuels using Figure 7.9.

FIGURE 7.9: CURVE SHOWING PERCENT EFFICIENCY IMPROVEMENT PER EVERY PERCENT REDUCTION IN EXCESS AIR

[Valid for estimating efficiency improvements on typical natural gas, #2 through #6 oils and coal fuels. (9)]

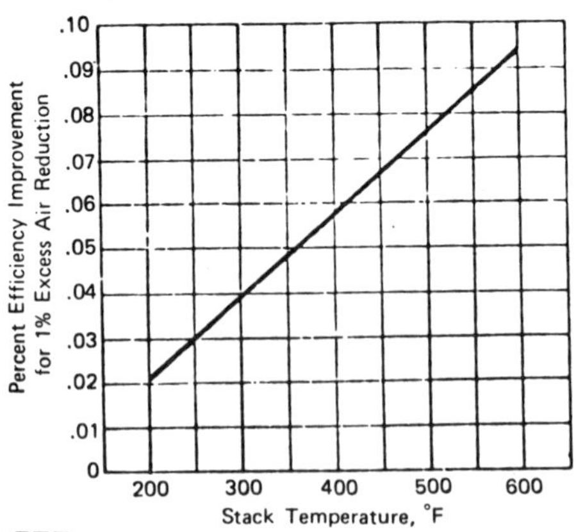

Source: PB-262 577

TABLE 7.1: DIFFERENCES IN PERFORMANCE OF CONTROL SYSTEMS (8)

Factors Affecting Accuracy of Fuel/Air Ratio Control	..... Potential Control System Error, Percent Excess Combustion Air......												Effective Corrective Action
	Fixed Positioning		Positioning with Operator Trim		Pressure Ratio		Fuel and Air Metering		Cross-Limited Metering		Oxygen Correction		
	Min.	Max.	Min.	Max.	Min.	Max.	Min.	Max.	Min.	Max.	Min.	Max.	
Fuel valve and positioning linkage	1	4	1	4	–	–	–	–	–	–	–	–	fuel metering
Upstream fuel pressure	1	3	1	3	–	–	–	–	–	–	–	–	fuel metering
Air damper and linkage	2	6	2	6	–	–	–	–	–	–	–	–	air metering
Fan performance	2	5	2	5	–	–	–	–	–	–	–	–	air metering
Alignment of burner characteristic with air differential	–	–	–	–	3	10	–	–	–	–	–	–	fuel and air metering
Burner wear and carbonization	4	10	4	10	4	10	–	–	–	–	–	–	fuel metering
Fuel Btu, specific gravity, and temperature variation	2	4	2	4	2	4	2	4	2	4	–	–	O_2 correction
Temperature, pressure and humidity errors in air metering	7	14	7	14	4	9	4	9	4	9	–	–	O_2 correction
Air/fuel control response lag	–	–	–	–	–	–	1	2	–	–	–	–	O_2 correction
Unaccounted for errors	½	1	½	1	½	1	½	1	½	1	–	–	O_2 correction
Control system accuracy	½	1	½	1	2	5	½	2	½	2	½	2	–

(continued)

TABLE 7.1: (continued)

Factors Affecting Accuracy of Fuel/Air Ratio Control	. . . . . Potential Control System Error, Percent Excess Combustion Air.												Effective Corrective Action
	Fixed Positioning		Positioning with Operator Trim		Pressure Ratio		Fuel and Air Metering		Cross-Limited Metering		Oxygen Correction		
	Min.	Max.	Min.	Max.	Min.	Max.	Min.	Max.	Min.	Max.	Min.	Max.	
Potential cumulative error	20	48	20	48	15.5	39	8	18	7	16	½	2	–
Probable cumulative error or additional excess air	34		34		27		13		11		1		–
Add for complete combustion (gas fuel)	43		43		36		22		20		9		–
Probable excess air required with operator trim (from combustion guide (gas fuel)	43		37		31		17		15		10		–
Add for complete combustion (oil fuel)	53		53		40		32		30		19		–
Probable excess air required with operator trim from combustion guide (oil fuel)	53		47		41		27		25		20		–
Type of combustion guide	–		steam flow/ airflow or oxygen analysis		steam flow/ airflow or oxygen analysis		–		oxygen analysis		–		–

Source: PB-262 577

The reduction in excess air described above will result in a 3% benefit in fuel savings for a unit with 600°F exit gas temperature. These values will be augmented by the reductions in stack gas temperature that accompanies the lower excess air flows. For the example given above, an additional 2% reduction in fuel savings will be realized. The gross annual savings can be estimated using the nomograph shown in Figure 7.2. Units equipped with economizers and/or air preheaters providing lower stack gas temperatures will have less of a savings.

Costs

Approximate costs for the six types of control systems to be described are given in Table 7.2 as per manufacturers. These costs will be affected by the type of equipment, either electronic or pneumatic, and the degree of sophistication employed. A fully computerized system employing direct digital control may have an installation cost of $100,000 per boiler.

Oxygen analyzers employing zirconium or ceramic oxide cost approximately $3400. These units have demonstrated very low maintenance requirements.

TABLE 7.2: APPROXIMATE COST OF BOILER COMBUSTION CONTROL SYSTEMS (One burner)

	Installed Cost, $ (1975)
Fixed Positioning	23,000
Positioning with Operator Trim	26,000
Pressure Ratio	31,000
Fuel and Air Metering	33,000
Cross Limited Metering	37,000
Oxygen Correction	43,000

Source: PB-262 577

Combustion Control Systems

The six basic combustion control systems employing a single fuel are described below. For further information on these systems, multiple fuel installations or other variations, please consult the references or manufacturers listed in the Appendix.

Fixed Positioning: A simplified fixed positioning control (jackshaft) system has been extensively applied to industrial boilers based on minimum control system costs. A single actuator moves both the fuel and air control devices through a mechanical linkage to a preset position in response to a change in the steam pressure. The system is calibrated by adjusting the linkage for proper fan damper travel and the cam on the fuel valve to produce the desired fuel/air ratio over the load range. A manual control override of the main actuator is usually provided.

Since the system merely positions fuel valve and air damper openings, it cannot compensate for changes in fuel or air density, fuel supply pressure or heating valve. Wear in the fuel valve orifice, burner tips and fan dampers will also alter fuel/air ratios. This type of control system results in considerable variations of excess air depending on the particular conditions existing at a given time. As a result, nominally high excess air levels must be preset to avoid going below the minimum excess air requirements. In actual operation, excess air will vary over a considerable range as indicated in Table 7.1.

Parallel Positioning with Operator Trim: A lower margin of operating excess air can be achieved by substituting a pneumatic or electronic positioning system which reduces the variations caused by mechanical linkages.

Separate actuators for the fuel valve and fan damper acting in parallel from a single steam pressure controller can be added. In addition, individual manual bias control of the fuel or air input can be used to adjust the fuel/air ratio. This system requires a combustion guide in the form of a steam flow/air flow or oxygen analyzer to assist the operator in positioning the excess air flow.

The parallel positioning control is the most widely used combustion control system for units below 100,000 pounds per hour (10). By adjusting the fuel/air ratio, compensations for variations in fuel characteristics, combustion conditions or control system equipment can be made.

Pressure Ratio: Use of the fuel pressure at the burner and the windbox to furnace pressure differential as indications of fuel and air flows are the most elementary of the metering type systems. A basic parallel pneumatic or electronic system is used with the windbox to furnace/burner pressure ratio employed to trim the fuel and air flows. Manual bias control to change the ratio of pressures and excess air flows is normally provided incorporating a combustion guide.

This system eliminates inaccuracies due to fuel pressure and fan discharge variations but it requires that the pressure-flow relationships for the fuel and air have similar characteristics. Alignment problems often exist causing inaccuracies. Variations in fuel heating value and fuel or air densities cannot be reconciled.

Fuel and Air Metering: If the control system is further refined to include actual metering of fuel and air, additional error sources can be eliminated, further reducing the excess air levels. Several types of flow metering devices are available; however, there are some limitations as to the types of fuels that can be accurately metered.

Several combinations of components can be used in a metering control system. Feedback loops are employed to allow the fuel and air to self-correct flows to meet system demands. Additional air is normally required to allow for the differences in speed of response between the fuel and air flow loops.

Cross-Limited Metering: An additional refinement to the metered system is the cross-limited metering system that limits the change in fuel flow through control logic to available air flow at all times. The quantity of air flow is also tied to the existing fuel flow and must be equal to or greater than fuel flow. Required excess air levels are thereby slightly reduced.

A common form of the cross-limited metering system is a pneumatic or electric parallel system using the steam pressure as the master controller. A manual override of the fuel/air ratio is provided to trim excess air levels.

Cross-Limited Metering with O_2 Correction: The effects on excess air of variations in fuel heating value and combustion air conditions can be eliminated using a continuous monitored flue gas O_2 level to trim the fuel/air ratio. The unit can thereby be operated at a set O_2 level resulting in minimum fuel consumption.

Combustion Control Equipment

Control systems hardware and instrumentation are either pneumatic or electronic. Considerations that must be included in the selection of equipment are

- capital costs–pneumatic systems cost 20% less than electric systems, especially final control elements

- installation costs–electric generally less expensive
- instrument air costs on pneumatic systems
- compatibility
- maintenance factors

Electronic components may be either analog or direct digital systems. Analog components are springs, diaphrams, cams, etc. that simulate mathematical functions used to determine the response to an incoming signal. Direct digital sytems employ a calculator to determine the desired response and position actuators. The same considerations mentioned above with respect to costs, compatibility and maintenance would again apply to the selection of a system.

Oxygen Analyzers: Oxygen analyzers are the accepted means of determining excess air levels. Recent developments have improved the old style sampling systems with analyzers using zirconium or ceramic oxide sensors so that little, if any, pretreatment of the gas sample is required. Sensors can be located directly in the flue gas stream or on the flue gas ducts connected directly to sampling probes. These systems have increased sensitivity especially in the lower excess O_2 levels and have significantly reduced sample cleanliness and maintenance requirements.

Instrumentation: Table 7.3 (10) presents the minimum instrumentation requirements to allow safe control of the boiler and provide a sufficient record of operation. Many additions to this list are possible and are subject to costs, legal requirements and personal preference of the plant operators.

Steam pressure, temperature, and flow along with air and fuel flow provide information on the operation of the control system in maintaining desired boiler operation at various loads. Necessary information of the boiler water supply is provided by drum level and feedwater flow indicators. Draft gauges can indicate plugging of the boiler, economizer, or air heater passages with fly ash. Gas and air temperature measurements also indicate the need for sootblower operation. Annunciators are required to warn the operator of hazardous operating conditions.

Air Pollution Emissions: Use of more sophisticated combustion controls to reduce excess air requirements will result in a reduction in NO_X emission levels.

TABLE 7.3: INSTRUMENT RECOMMENDATIONS

	Indicator	Recorder	Integrator
Boiler outlet pressure	R	R	
Superheater outlet steam temp		R	
Steam flow		R	X
Total fuel flow		R	
Total air flow		R	
Individual fuel flow		X	X
O_2 analyzer		X	
Combustibles analyzer		X	
Drum level	R	R	
Feedwater flow		X	X
Draft gauges	R		
Air and gas temp	R	X	
Feedwater temp	R		
Annunciator	R		

R = Required
X = Optional

Source: PB-262 577

OIL AND GAS BURNERS AND SUPPLY SYSTEMS

This section describes the available auxiliary equipment that can be employed to improve burner performance for oil and gas fired systems. For methods of improving burner performance other than by means of auxiliary equipment, see Chapter 3.

Principle of Operation

The preceding section discussed the impact of excess air levels on boiler efficiency. The additional air results primarily from excess air at the burner required to provide complete combustion of the fuel. Current design trends have employed modified fuel/air mixing and flow patterns to minimize excess air requirements. Associated reductions in NO_X and particulate emissions have also been demonstrated.

Low excess air (LEA) burners originally developed in Europe are currently being applied to industrial steam generating units in the United States. These combustion systems employ improved oil fuel atomization and controlled air fuel mixing to achieve extremely low excess air (3-5%) operating levels. Many of these systems were initially developed to reduce air pollution levels and can therefore offer this additional benefit.

Some restrictions, however, have been encountered in the application of LEA burners to industrial boilers especially on a retrofit basis. These problems have been primarily associated with the compatibility of new flame shapes and heat release rates with existing furnace volume and design. A very sophisticated combustion control sytem employing automatic O_2 correction is required with an LEA burner to fully utilize its low excess air capability and prevent small excursions in excess air which might lead to high combustibles and potentially unsafe conditions.

Conventional packaged burner systems that come complete with fans and combustion controls are also available. While these systems are not as sophisticated as those described above, they can result in substantial reductions in required operating excess air levels where existing burners require high excess air due to outdated design, wear, deterioration, etc. They also have the advantage of reduced maintenance, convenient replacement and are readily adaptable to nearly all industrial boiler configurations.

Other equipment that can benefit boiler efficiency includes special oil fuel atomization and viscosity control systems. Proper pretreatment and atomization of the oil fuel are vital to reduced excess air operation.

Performance

The performance benefits using improved combustion equipment are determined by the resulting reduction in operating excess air levels.

Several manufacturers of LEA burners have claimed operating excess air levels as low as 2%. The resulting efficiency improvement over normal operation can be estimated using Figure 7.9. It has been questioned however, whether these excess air levels can be maintained throughout the entire load range and over prolonged operation especially on multiple burner systems. Combustion control systems use an average flue gas excess O_2 level measured at the boiler outlet to adjust all the burners as a single unit. Variations in air or fuel flow between burners due to normal oil tip wear, gradual dissimilarities in air damper adjustments, etc. may result in the inability to maintain extremely low excess air levels without extensive maintenance requirements. More realistically attainable levels may prove to be approximately 5% excess air.

Manufacturers of packaged burner systems guarantee an operating level of 10% excess air for both oil and gas fuels. Again the associated efficiency improvement over normal operation can be estimated using Figure 7.9.

Costs

Equipment costs for a LEA burner system range from $5,000 to $50,000 for the size of boilers considered here. Required upgrading of the combustion control systems will involve additional expenses as discussed in the previous section. Packaged burners include the combustion control system and vary from $10,000 to $45,000. The number of burners, boiler configuration and required modifications to existing combustion air and fuel supplies will affect these price ranges. Several options are available that can significantly increase these costs including upgraded fuel handling system, safety controls, stack monitoring, etc. Installation costs will be 50 to 100% of the equipment costs.

Installed costs for retrofitting oil burners with improved atomizers can be as high as $10,000 per unit depending on the type and number of burners. Sophisticated viscometer systems are approximately $3,000.

Types of Oil and Gas Burners

LEA burners generally resemble conventional burners. Modifications in fuel injection and air flow patterns to achieve optimum mixing and flame patterns permit very low excess air while maintaining complete combustion. As with conventional burners, the minimum excess air levels are indicated by excessive levels of CO or hydrocarbons on gas fuel and smoke on oil fuel.

Current design trends have emphasized the use of primary and secondary combustion zones within the flame to minimize NO_X formation. The primary zone nearest the burner is maintained fuel rich with fuel and air mixing characteristics designed to provide stable flame holding. Swirl or rotation in the air stream or in certain cases turbulence from deflectors provide the necessary flow pattern. Additional air required for complete combustion is provided in a secondary combustion zone surrounding the primary zone.

The combustion of oil fuels involves an additional aspect not present in the combustion of gaseous fuels. Oil fuel must be atomized prior to mixing with the combustion air. This is accomplished by dispersing the oil into a fine droplet spray that evaporates from the heat of the flame. The fineness of the atomization determines the evaporation rate of the oil and will vary with the viscosity of the oil entering the atomizer.

LEA burners develop the fuel spray and air flow patterns to achieve both complete mixing and a stable flame pattern while requiring less overall excess air than conventional burners. Fuel sprays are oriented or swirled in several fashions to direct the oil vapors into the air flow. Various air flow patterns are employed to provide a controlled mixing pattern for complete combustion with minimum excess air requirements. This flow pattern holds the mixture in proper position in front of the atomizer and provides for vaporization and ignition of the oil droplets by locally recirculating exhaust products and by radiation from the flame.

Numerous factors must be considered in the selection of a new burner system. Flame shape and heat release rates must be compatible with furnace conditions. Use of a system that can employ existing fan and fuel handling capabilities will significantly reduce costs. Adequate turndown capabilities and firing rate response for load modulations is important. Flame scanners and other safety equipment and instrumentation may be useful to improve reliability. A comprehensive engineering evaluation and consultation with the boiler manufacturer is recommended prior to the selection of new burner equipment.

Application Considerations

Atomizing Systems: Efficient combustion of a liquid fuel requires rapid vaporization of the fuel which is facilitated by atomizing the fuel to the smallest possible droplet size.

Many new concepts for improved fuel oil atomization are currently available. One system employs sound waves to "sonically atomize" the fuel to a fine liquid spray. Another system uses water emulsion in the oil that flashes to steam at the burner nozzle and shatters the oil into small particles. The manufacturers of these techniques generally claim improved turndown ratios and as much as a 5% savings in fuel when compared to the conventional steam, air, mechanical or pressure atomizers currently utilized in industrial boiler burners.

When evaluating new atomizing systems it should be recognized that these will impact boiler efficiency primarily by permitting operation at lower excess air levels thereby minimizing the "dry gas losses." For retrofit applications these systems are therefore directed to existing burners which are of inadequate design to operate with low excess air.

Viscosity Control: Viscosity is a measurement of the relative ease by which a liquid will flow when subjected to a pressure differential. Viscosity is important to burner performance in the direct relationship it has with atomization. Droplet size, spray cone angle, pressure drop, and soot formation in the burner are all affected by the viscosity of the oil. Proper atomization is basic to achieving complete combustion with low excess air.

For optimum operation, all oil burners depend on an accurate control of the oil's viscosity. Common pactice is to control oil temperature using a known relationship between temperature and viscosity. Recent viscometer developments have made direct viscosity control possible. Manufacturers of such instruments claim that direct viscosity control is now necessary due to the wide variations in oils currently on the market.

Air Pollution Emissions: One of the primary factors in the development of low excess air burners has been the desire to reduce NO_X and particulate emission levels. Manufacturers claim that up to a 20% reduction in NO_X emissions can be achieved using a LEA burner with corresponding reductions in particulate levels.

Lower NO_X emission levels can be expected with low excess air operation on any industrial burner. Typically, NO_X levels are reduced 20-100 ppm for every 1.0% decrease in excess O_2 (6).

WALL AND SOOT BLOWERS

Principle of Operation

Wall and soot blowers are employed to remove deposits from heat transfer surfaces in the boiler. Removal of slag deposits from the furnace walls of a coal-fired unit is done using wall blowers. Soot blowers are employed to remove fly ash and soot deposits from the convective passes of the boiler. If allowed to accumulate, these deposits would retard heat transfer and may eventually result in clogging the boiler.

These problems are related to oil and coal fired systems. The properties of the fuel, the characteristics of the firing system, and the resulting temperature distribution in the boiler have an effect on the type and accumulation rates of the deposits. Analyses of these factors are required prior to selection and installation of a soot blowing system. Retrofit applications are often required with changes in fuel supply or when existing equipment is deemed inadequate due to excessive stack gas temperatures or tube failures.

Performance

Manufacturers claim that proper soot blowing can increase unit efficiency by up to 1%.

Costs

The cost of retrofitting a soot blower system to an existing unit will be dependent on the availability of an adequate cleaning medium (either air or steam) and the size and number of units required. An approximate installed cost for a simple system (i.e., three blowers) would be $15,000-50,000 depending on the size of the unit.

Physical Description and Application

Soot blowers can be either fixed or retractable and use either air or steam as the blowing media. Fixed position systems are employed in low temperature regions whereas retractable systems, known as lances, are generally used in high temperature regions. Special nozzles and arrangements are employed including rotating lance blowers as illustrated in Figure 7.10. Wall blowers are generally short retractable lances that penetrate the boiler furnace wall and rotate 360 degrees.

The availability of steam and compressed air and the cost of installation and maintenance of the supply system will determine the selection of the cleaning medium. The choice of the cleaning medium will also depend to a degree on the characteristics of the deposits. Slags at the cold end of their "plastic" temperature range (liquid and solid mixture) can be chilled (solidified) and removed by air or steam blowers. Deposits that will remain plastic over a large range in temperature (greater than 50°F) may require a water spray to solidify the deposits for easier removal.

Air Pollution Emissions: Particulate emissions may temporarily increase when soot blowers are operated.

FIGURE 7.10: SOOT BLOWER ARRANGEMENTS*

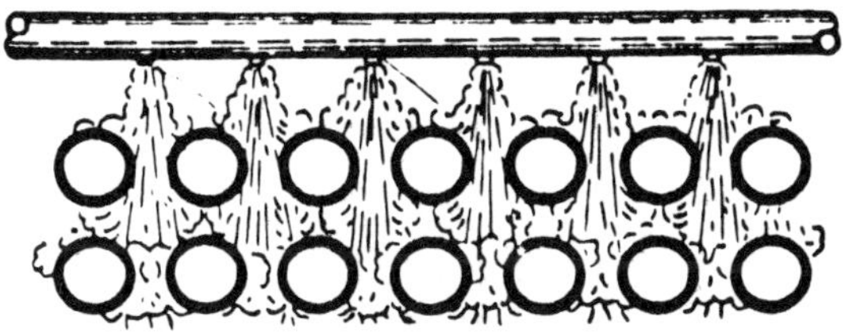

Stationary Soot Blowers employing straight nozzles clean banks of in-line tubes. With staggered tubes, the nozzles are offset for improved penetration.

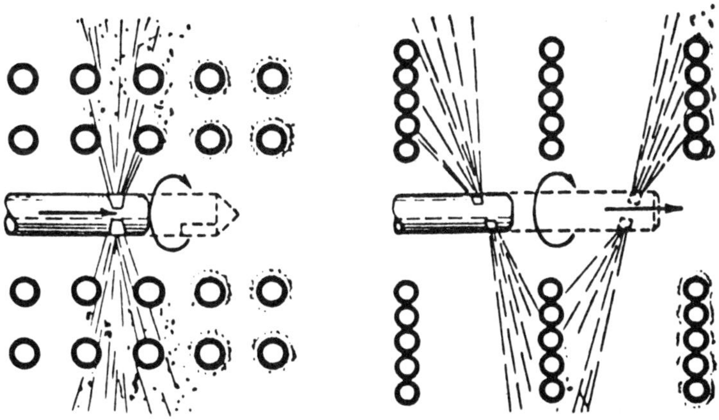

Retractable Blower Lance, driven by two motors, traverses boiler at one speed, retracts at twice that speed. Result: about a 25% saving in cleaning time and in blowing medium. Speed of lance rotation is held throughout the cycle.

*Reprinted with permission from *Power* magazine, June 1964.

Source: PB-262 577

INSULATION

Principle of Operation

Insulation is any material that is employed to restrict the transfer of heat energy. It can generally be categorized as either mass or reflective type depending on whether it is aimed at reducing conductive or radiative heat transmission, respectively.

The overall thermal resistance of the mass type insulation depends on both its composition and physical structure. Materials are effective that have a low thermal conductivity but more importantly contain numerous small voids filled with air or gas. These voids provide the primary resistance to conductive heat transfer. The effectiveness of the insulation is also determined by the thickness or quantity of material used.

Reflective type insulation, generally consisting of smooth metal surfaces, reduces the loss of heat energy by radiation. This type of insulation is often used as a covering for mass type insulation and can also be employed in layers separated by air pockets to replace mass insulation. This application offers several advantages; however, consideration must be given to heat loss by convection and conduction through and between the metal surfaces.

Insulation provides other benefits besides reduced heat losses including controlled surface temperatures for comfort and safety, structural strength, reduced noise, and fire protection.

Performance

Properly applied insulation can result in large savings in energy losses depending on type, thickness (mass type only), and condition of the existing insulation. Figure 7.11 presents the approximate total rate of heat energy loss by radiation and convection from a bare flat surface based on the temperature difference between the surface and still air.

FIGURE 7.11: HEAT ENERGY LOSS FROM BARE SURFACE*

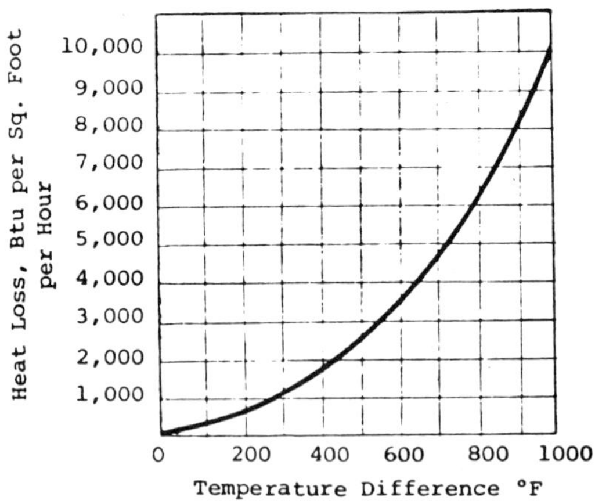

*Reprinted with permission from *Power* magazine, August 1974.

Source: PB-262 577

Bare surface temperatures in steam generating units range from saturation temperature on exposed tube surfaces to air and gas temperatures on duct surfaces. Approximate radiation losses from furnace walls have been developd by the ABMA (11). They show that radiation losses tend to increase with decreasing load and can be as high as 7% for small units or larger units operating at reduced loads.

More precise determination of energy losses can be made based upon existing insulation operating temperature, air flow over the surface, surrounding conditions and surface orientation (horizontal or vertical). An engineering analysis of existing conditions and potential energy savings should be conducted prior to application of insulating material.

Costs

Procedures have been developed by the Thermal Insulation Manufacturers Association to determine the optimum insulation thickness for various applications based on

- fuel costs
- operating temperatures
- insulation type
- depreciation period of plant and insulation
- capital investment

Types of Insulation and Application

Mass Type: The selection of an insulating material is based on several factors other than composition. As mentioned, the numerous voids contained in the material structure provide the primary resistance to heat transmission by conduction. Mass type insulation must be protected from damage by moisture absorption or excessive compression as these will reduce the insulating value of the material.

Common insulating materials include calcium silicate, cellular glass and diatomaceous silica. Insulation made of 85% magnesia and asbestos that was once extensively used in high pressure steam generators is no longer made and has been replaced by glass fiber materials. Temperature limitations sometimes require the use of dual insulating materials for effective insulation of high temperature surfaces.

Stiffeners, metal mesh, or protective jackets are employed to increase the strength and dimensional stability of the insulation. Protection against fire is an important consideration both with regard to the flammability of the insulating material and protection of the insulant from external fire.

Reflective Type: Reflective type insulation generally consists of smooth metal surfaces that have a low emissivity and so retard the loss of heat energy by radiation to cooler surroundings. Table 7.4 (2) presents the emissivity of various materials that represent the ratio of the metal's radiating power to the maximum radiation rate at that temperature. In comparison, ordinary paint or mass type insulation covering have emissivities of approximately 0.85 to 0.95.

TABLE 7.4: EMISSIVITY OF METALLIC SURFACES

Metal	Polished Metal Surfaces Temperatures 70°F	Polished Metal Surfaces Temperatures 1000°F	Oxidized Surfaces Temperatures Below 1500°F
Aluminum	0.05	0.075	0.10 - 0.20
Copper	0.04	0.08	0.55 - 0.75
Gold	0.03	0.05	
Iron, cast or wrought	0.20	0.25	0.60 - 0.90
Monel metal	0.07	0.10	0.40 - 0.50
Nickel	0.06	0.10	0.40 - 0.60
Platinum	0.036	0.10	
Silver	0.20	0.25	
Steel	0.20	0.25	0.60 - 0.90
Tin	0.08	--	

Source: PB-266-577

Use of a single layer of reflective insulation does not impede the transfer of energy by convection or conduction; however, multiple layers separated by air pockets can be employed to form an insulating barrier that is more effective than mass type insulation in certain applications. Convection losses in the air separating the layers can be reduced by seals and transverse metal barriers that restrict air movement due to buoyancy forces. Conduction losses at contact points must also be considered.

Reflective type insulation offers the following advantages over mass type insulation: lightweight, resists moisture damage, resists fire damage, easier to decontaminate, and adaptable to several configurations.

The greater susceptibility to corrosive damage and heat energy losses by convection and conduction are the primary disadvantages.

Air Pollution Emissions: Lower emissions will result on a lb/lb fuel basis due to higher unit thermal efficiency and therefore less fuel consumption.

AUXILIARY EQUIPMENT vs A NEW UNIT

Purchase of a New Steam Generating Unit

As previously mentioned, the cost of adding auxiliary equipment to improve unit performance on an existing boiler may justify the purchase of an entirely new system. This is especially true of an older unit with a limited expected useful lifetime or a unit with severe space or operating constraints that require additional expensive modifications. Other factors may prohibit the use of various equipment or techniques and limit the boiler's efficiency improvement potential.

Once the purchase of a new system has shown to be cost effective, the selection of the most efficient unit incorporating the optimum use of the auxiliary equipment discussed in the previous sections becomes all important. Several manufacturers representing various boiler configurations should be consulted to select the best system on the basis of initial, operating, and maintenance costs. It is generally advised to select standard units available commercially, as special design specifications will add significantly to the costs of purchasing the unit.

When purchasing a new boiler it is very advisable to compare boiler efficiencies of prospective units since even a few tenths percent efficiency advantage can lead to substantial savings in fuel over the life of the boiler. It is common for boiler manufacturers to offer a performance guarantee which will contain at least the guaranteed minimum boiler efficiency at maximum boiler capacity. Predicted performance at intermediate loads may also be provided, especially for larger boilers.

Ideally, the complete efficiency versus load characteristics of the boiler would be available to enable efficiency comparisons at the particular range of loads anticipated for the boiler. Unfortunately, this information is very seldom available and the efficiency characteristics of the boiler are determined only after installation and start up.

The excess air characteristics of the burner (i.e., excess O_2 versus load profile) are very important in determining the efficiency vs load curve of the boiler. This is especially true for boilers without waste heat recovery where stack temperatures are higher and the efficiency is consequently more sensitive to the quantity of excess air used.

Therefore it may be desirable to avoid burner or furnace designs which require high excess air. Increased excess air requirements at lower loads are quite common but some combustion control systems and burner designs are available which permit very low excess air levels over the entire turndown range of the boiler. Choosing this type of system would be very attractive especially when the boiler will be operated at lower loads a large fraction of the time.

Other factors not directly connected to boiler costs and efficiency must be considered. Changes in steam requirements or operating conditions may warrant a completely different type, size or firing mode than previously employed in the existing system. Space limitations or steam demand flexibility can make the use of several smaller units more desirable than one large unit.

Consideration should also be given to the "quick steaming" boilers that can come to operating temperature and pressure in a few minutes. These units are very compact based on a once-through design that eliminates the larger water jackets of comparable capacity firetube designs. The steady state operating efficiency of these units is comparable to a conventional boiler at rated loads but they can produce significant fuel savings in applications where the boiler is used for only one shift a day.

A conventional firetube boiler radiates a considerable quantity of heat as the water jacket cools after shutdown. This heat is wasted and must be replaced when the boiler is refired for the next day of use. Since the once-through units have a very small water volume, they can produce steam in minutes and save the extra fuel a conventional boiler would use for startup. Using the boiler manufacturers' example of a 10 hour day and a 40 minute startup period at 81% efficiency for the conventional drum boiler, the once-through steam generator would use 17% less fuel. Some quick steaming boilers also have smaller casing surface area (compared to conventional boilers with equivalent capacity) which reduces radiation heat losses and leads to higher boiler efficiencies, especially at reduced firing rates.

Numerous other options are available to the boiler purchaser. New air pollution constraints and changing fuel supplies may alter the fuels employed. This will strongly influence the method of firing and to a certain extent the configuration of the unit. In all cases, a complete engineering and financial evaluation of possible retrofit applications to the present system and the purchase of a new system is justified prior to the selection and installation of a new steam generating unit.

REFERENCES

(1) Shields, C.D., *Boilers, Types, Characteristics and Functions,* McGraw-Hill, 1961.

(2) Salisbury, J.K., *Kent's Mechanical Engineers' Handbook,* John Wiley and Sons, Inc., 1967.

(3) Literature from the C.-E. Air Preheater Company, Wellsville, NY.

(4) Babcock & Wilcox Company, *Steam - Its Generation and Use,* 38th Edition, 1972.

(5) Sales Literature from Hughes Electron Dynamics Division, Torrance, CA.

(6) Cato, G.A., et al., *Field Testing: Application of Combustion Modifications to Control Pollutant Emissions from Industrial Boilers - Phase II,* KVB, Inc. Report to the U.S. Environmental Protection Agency, No. EPA 600/2-76-086a, April 1976.

(7) Clark, Norman, *Higher Efficiency Through Lower Stack Temperatures,* The C.-E. Air Preheater Company.

(8) Dukelow, S.G. "Energy Conservation Through Instrumentation" Lecture to George Washington University, April 1975.

(9) KVB, Inc. *Industrial Boiler Operating Guidelines for Improved Efficiency and Reduced Stack Emissions,* U.S. Environmental Protection Agency, Contract 68-02-1074, and the Federal Energy Administration No. C-04-50085-00.

(10) Delin, M.A., "Control Requirements for Industrial Boilers," Presented at the American Power Conference, May 1973.

(11) ASME Power Test Code 4.1, 1964.

Boiler Maintenance Practices

The material in this chapter was excerpted from reports by KVB, Inc. (PB-262 577 and PB-264 543) and by Engineering Extension Service, Auburn University (PB-265 713).

The industrial boiler user has a number of options for improving efficiency and reducing fuel consumption without investing large amounts of capital. Some of the most cost-effective approaches require simply taking steps to restore the boiler to its original design operating efficiency and initiating preventive maintenance procedures to insure that a high level of efficiency is maintained. There are, of course, many older boilers in operation where low design efficiency and excessive maintenance expenses would show complete boiler replacement to be the correct economic choice. Modifications to boiler operating pactices through reduced operating pressures, blowdown control and load management can benefit unit performance.

GENERAL MAINTENANCE PROCEDURES

Historically, preventive maintenance in an industrial boiler plant has been performed to provide for personnel safety, and to prevent damage to equipment which could result in costly repairs and loss in productive capacity. Additional maintenance requirements have been generated more recently by pollution control regulations and the Occupational Safety and Health Act.

Prior to the energy crisis in 1973, preventive maintenance directed specifically to maintaining boiler efficiency was the exception rather than the rule. Now, rising fuel costs have placed increasing emphasis on conscientious maintenance which is necessary in preserving high boiler efficiencies. These maintenance practices are in many cases easily justified on an economic basis.

Efficiency related boiler maintenance is directed towards correcting any condition that resulted in an increase in fuel consumption to generate a constant quantity of steam. Thus, at a given boiler load, any condition which produces an increase in: 1) stack gas temperature, 2) stack gas flow, 3) combustible content in ash or flue gas, 4) convection or radiation losses from the boiler exterior, ductwork or piping, or 5) blowdown quantities, is considered an efficiency related maintenance item. These maintenance items may be primarily efficiency related or primarily safety/capability related; the distinction is usually a matter of preference or degree. In any event, attention to efficiency related maintenance items can also generally forestall more serious consequences which could result in damage to equipment or injury to personnel.

Boiler Tune-up

One of the more effective means of improving and maintaining boiler operating efficiency is a boiler "tune-up." This preventive maintenance item is one of the most direct approaches to fuel conservation through efficiency improvement.

The primary objective in a tune-up is to achieve efficient combustion with a controlled amount of excess air. Operating with the lowest practical excess air will minimize efficiency losses by reducing the quantity of unneeded air which is heated to the stack temperature then not utilized. The associated reduction in stack gas temperature and power consumption by forced draft and induced draft fans are additional benefits. The actual improvement in boiler efficiency with lower excess air depends on the initial stack temperature and excess air at the boiler exit. A given change in excess air will have a greater effect when stack temperatures are high. For example, a reduction in excess air of 10% (i.e., from 20 to 10% or 100 to 90% excess air) will produce a 0.9% improvement in efficiency when stack temperatures are on the order of 600°F.

However, the same change in excess air when stack temperatures are near 200°F will improve boiler efficiency by only 0.2%. These values are not appreciably affected by the type of fuel (natural gas, oils or coal). Cases have been documented (1) where reductions in excess air by much more than 10% were possible leading to efficiency improvement of several percent.

Proper operation of the boiler combustion control system is essential for maintaining high boiler operating efficiencies and satisfactory excess air levels. Its main purpose is to provide the correct quantities of fuel and air at the burner to satisfy a varying demand for steam generation. Although it is important that the excess air delivered to the burner be kept to a minimum over the boiler operating range, it generally is not practical to operate precisely at the point of maximum efficiency. This optimum typically occurs at the "threshold" of combustible or smoke formation and may result in unacceptable stack conditions. For the majority of boilers, it is necessary to maintain a margin of excess air above the minimum or threshold level to accommodate variations in fuel properties and ambient conditions, nonrepeatability of control settings, normal deterioration of control parts and rapid changes in firing rate.

Proper adjustment and maintenance of the burning equipment are essential to good combustion control. Burner parameters such as air register position, location of the diffuser or position of coal spreaders can affect the excess air requirements. Fuel oil temperature and atomizing pressures, pulverizer performance (coal fineness), primary air temperature, and other operating factors also have an impact.

To assure reliable, safe, and efficient boiler performance, manufacturers of boiler and burner equipment recommend periodic inspections and tune-ups. Thorough tune-ups are commonly recommended to be performed on an annual basis but many operators also prefer to conduct quick boiler efficiency checks much more often, sometimes on a daily or weekly basis.

In this way, efficiency problems can be detected early, before large fuel wastage occurs or expensive maintenance is required. Operators can perform their own tune-ups and efficiency audits, or boiler tune-up services are also available from most major manufacturers of boiler and burner systems, some local utilities, and engineering consulting firms. Since tune-up crews usually charge an hourly or daily rate plus expenses, costs for these services will depend on the extent of the tune-up work required.

A minimal tune-up should include a verification of automatic fuel and air control operation over their operating range. Visual furnace observations and stack measurements of excess O_2, CO, CO_2 and temperature are essential elements in this type of tune-up. A typical tune-up performed on a 13,000 lb/hr watertube boiler during an FEA field testing program (2) required one day and cost about $300.

A more extensive tune-up might also involve a boiler shutdown and a thorough inspection of burner parts, dampers, fuel valves and regulators, refractory, furnace tubes and instrumentation. In many cases the more critical maintenance, instrument calibration and repairs can be made on-site during a brief shutdown. Major repairs may require an extended shutdown at a later, more convenient date. A list of additional items requiring attention at that time should be furnished. For boiler operators desiring a basic efficiency check of their boiler, the local natural gas supply company may provide this service at little or no charge to customers and may also offer some assistance in adjusting burner controls for peak efficiency.

For a complete combustion control tune-up and inspection, the tune-up crew will prefer to have total control of boiler firing rate to verify control settings over the full turn-down range. Be prepared to accommodate the fluctuations in steam flow and/or pressure.

In some cases it may be necessary to install a baffled steam discharge stack when these fluctuations cannot be tolerated in the steam supply system or plant processes. When stack measurements are being made (O_2, CO, CO_2 and temperature), it is important to request a verification of stack gas uniformity (i.e., a traverse) to assure representative readings. Flue gas samples are usually extracted by means of a single point probe and any stratifications or gradients in the flue gas ducting can lead to erroneous measurements.

It is important that excess air not be reduced at the expense of excessive combustibles (unburned fuel, carbon carryover, CO, etc.) since these can represent significant efficiency losses. More than 400 ppm carbon monoxide (CO) in the stack gases is generally not acceptable. Finally, it may be worthwhile for future reference to inquire about the margin in excess air used by the tune-up crew to accommodate the various uncontrollable factors mentioned previously.

For specialized combustion or operating problems not routinely solved by plant operating or engineering personnel, assistance from outside consulting or engineering firms should also be considered. Some firms have extensive experience in specific problem areas such as tube corrosion and wastage, excessive carbon carryover, precipitator and mill fires, water quality control, instrumentation malfunctions, and air pollution restrictions.

Boiler Tube Cleanliness

Tube deposits and fouling on the external tube surfaces of a watertube boiler or similar gas side conditions in the gas tubes of a firetube boiler will inhibit the absorption of heat in the boiler and lead to lower efficiencies. This condition will be reflected in high flue gas temperatures when compared to "clean" conditions at a similar firing rate and boiler excess air. The resulting loss in boiler efficiency can be closely estimated on the basis that a 1% efficiency loss occurs with every 40°F increase in stack temperature.

It should be mentioned that water side deposits resulting from inadequate water treatment would also eventually lead to higher stack temperatures and lower efficiencies. However, tube failures due to overheating would generally occur before any substantial efficiency losses are evident.

Poor firing conditions at the burner can be the major cause of gas side tube deposit problems. Insufficient burner excess air or improper burner adjustment and maintenance can lead to excessive carbon and soot formation in the furnace which builds up on furnace wall tubes and convective tube banks. Other factors to be considered might be fuel oil burning temperature, coal ash fouling properties, pulverizer performance and primary air temperature. A boiler inspection will generally indicate whether soot blowers are performing as designed and are properly located for effective tube cleaning. If existing soot blowers are functioning properly but are inadequate for present fuels, repositioning of soot blowers and/or the installation of new soot blower designs might be considered.

Stack temperature measurements are an easy and effective means for monitoring boiler tube cleanliness conditions. Stack temperatures should be periodically compared to values obtained during start-up or following a boiler tube wash to determine any deviations from "clean" baseline temperatures.

Since stack temperature usually increases with firing rate and excess air, these comparisons should be performed at similar boiler operating conditions. In the absence of any reference temperatures, it is normally expected that stack temperatures will be about 150° to 200°F above the saturated steam temperature at high firing rates in a saturated steam boiler. Naturally, this does not apply to boilers equipped with economizers or air preheaters. The boiler manufacturer should be able to supply a normal range of stack temperatures for the particular boiler design. When higher stack temperatures are measured, a boiler tube cleaning may be warranted.

Some boiler operators perform tube washes in-house but cleaning services are also available from commercial boiler repair and service companies. These companies also perform water side inspection and cleaning. Various cleaning methods are used including high-pressure water, air and steam jets to remove gas side deposits. Mechanical "shakers" and "knockers" are also used which vibrate off or dislodge the tube deposits for easy removal. The appropriate type of cleaning device will be dictated by the severity and composition of deposits, available access to tube surfaces, boiler construction (firetubes versus watertubes), age of the boiler, and other factors. The cost of an extensive tube cleaning will typically range from a few hundred dollars for a 10,000 lb/hr boiler to several thousand dollars for larger units. This expense is usually very easy to justify due to the quickly realized savings in fuel costs.

Determining Efficiency Related Maintenance Requirements

The implementation of an efficiency related maintenance program requires that performance deviations from optimum conditions be monitored and corrected either by operating or maintenance procedures. Many industrial boilers will require additional instrumentation for this purpose.

A basic tool for determining the need for maintenance to restore efficiency is the boiler operator's log. Daily trends in unit performance can be used to determine the deviation from design or optimum performance and establish the frequency for maintenance procedures. The boiler operator's log should contain the following information.

1) General data to establish unit output
   steam flow, pressure
   superheated steam temperature (if applicable)
   feedwater temperature

2) Firing system data
   fuel type (in multifuel boilers)
   fuel flow rate

oil or gas supply pressure
pressure at burners
fuel temperature
burner damper settings
windbox to furnace air-pressure differential
other special system data unique to particular installation

3) Air flow indication
stack gas O_2
optional - air flow pen, forced draft fan damper position, forced draft fan amperes

4) Flue gas and air temperatures
boiler outlet
economizer or air heater outlet gas
air temperature to air heater

5) Unburned combustible indication
CO measurement
stack appearance
flame appearance

6) Air and flue gas pressures
forced draft fan discharge
furnace
boiler outlet
economizer differential
air heater air and gas side differential

7) Unusual conditions
steam leaks
abnormal vibration or noise
equipment malfunctions

8) Blowdown operation

9) Sootblower operation.

Comparison of daily log data obtained with the unit in good operating condition may disclose the need for specific maintenance procedures. These observations should be made under very steady firing conditions. A burner and control system tune-up would be indicated if the unit smokes at previously acceptable O_2 levels. Higher than normal O_2 levels may indicate the need for control system adjustment or in the case of balanced draft units, may indicate excessive boiler setting leakage. A deficiency in air flow capacity may indicate excessive air heater seal leakage, plugged air heaters or other boiler and damper problems. The boiler operator's log will assist in assessing the problem.

Annual outages are also necessary for inspection of areas not accessible during operation to determine condition of boiler baffles, boiler cleanliness, air heater corrosion or seal condition, fan blade condition and damper operation.

The major areas of concern are burner-control system maintenance and boiler cleanliness, just described. Some less common but significant maintenance items from an efficiency standpoint are discussed in the following section.

Special Maintenance Items—Coal Fired Units

Many spreader stoker fired units incorporate fly ash reinjection from boiler hoppers or mechanical collectors to reduce carbon losses by reburning. Efficiency improvements of up to 5% are obtained by reinjection. Maintenance on this system is required to assure that the mechanical collector tubes are operating efficiently (no holes worn through collecting devices), the hoppers are being emptied and the conveying pipes are not eroded to the point of failure.

Stoker grate surfaces must be maintained to provide good air distribution and minimize sifting losses. Feeder devices must be kept in good repair to provide uniform fuel distribution.

In pulverized coal units the coal fineness should be checked regularly and the mills adjusted to provide the correct fineness. High carbon losses will result with poor grinding. Fans or exhausters handling pulverized coal/air mixtures are subject to erosion which can lead to reduced air flow which reduces mill capacity and can lead to fuel line plugging. Fan blades and grinding elements should be replaced as required to maintain performance. The frequency of replacement will depend on the abrasiveness of the coal being pulverized.

SYSTEMS AND EQUIPMENT

This section is a general review of maintenance items and procedures to be followed for the various systems within a boiler. A detailed description of sample record keeping is in the last section.

Fuel Supply System

Gas Firing: Since gas is a clean fuel and can be burned without any in-plant treatment, few problems are anticipated. The gas pressure regulator is the major component which might require attention.

Gas pressure on the downstream side of the regulator should be entered on an hourly log. Low pressure may indicate malfunction of regulator, plugging of upstream strainers or low pipeline pressure. High pressure or erratic pressure would indicate regulator malfunction. Deviation in gas pressure could influence fuel/air ratio depending on the type of control system.

Oil Firing: The fuel oil supply system consists of tanks, pumps, strainers and a pressure regulating valve to supply oil to the flow control valve at the proper pressure. In addition, tank heaters are required with heavy oil firing to provide the proper viscosity for pumping, and preheaters are required to provide the proper viscosity for atomizing. If the oil supply is extremely variable, viscosity control devices may also be included in the system.

With oils that do not require preheating, the components requiring attention are the strainers and oil pressure regulators. Heavy oil systems additionally require attention to tank heaters and preheaters.

Oil supply pressure and temperature should be entered on an hourly log as well as flow rate if metered. Low oil pressure may be caused by too high or too low temperature at the pump inlet, plugged strainers, coking in the preheater, or regulator malfunction. Preheater fouling or coking may also be indicated by inability to maintain desired fuel oil temperature.

Deviations in fuel oil pressure could influence fuel/air ratio depending on the type of control system. Viscosity control is important to providing proper fuel atomization.

Coal Firing: Most of the maintenance connected with the fuel supply system of stoker fired units is to keep the system functioning rather than efficiency related. With pulverized coal firing, the pulverizers and piping system performance can affect unit efficiency.

Hourly log readings should include pulverizer outlet temperature and pressure. Visual observation of the fires should be made to determine that fuel supply is reasonably uniform to all burners and that ignition point is normal. Monthly fineness checks should be made to determine that adequate fineness is being maintained. Annual inspection (or as recommended by the manufacturer) should be made to determine wear on grinding parts, classifiers and exhausters or fans. Wear on pulverized fuel pipes or riffle dividers should also be determined.

Poor grinding can increase carbon heat losses. Excessive grinding or worn pulverizers can increase power consumption. Low outlet temperature may indicate low pulverizer air flow, insufficient hot air, high moisture fuel, or wear on exhauster.

Controls and Instrumentation

Controls and instrumentation are the most critical areas in an efficiency-related maintenance program. The control function in maintaining the proper fuel/air ratio is essential to minimizing stack gas losses and unburned fuel loss. Accurate and reliable maintenance of boiler variables is also essential to monitoring boiler performance and analyzing equipment performance problems.

A variety of control systems are available. The components may be mechanical, pneumatic, or electronic, or combinations of various devices. Each type has certain weaknesses which require maintenance to assure satisfactory performance. Mechanical systems are subject to misalignment, binding, and wear which introduce errors in repeatability of settings. Pneumatic systems are subject to leakage or dirt in sensing and control lines. Electronic systems are subject to dirty contacts, insulation failure or circuit problems, and may be more sensitive to heat or vibration damage. The same types of problems are inherent in the various types of instruments used for performance monitoring.

The frequency of maintenance and calibration is best determined by plant operating and maintenance personnel based on records of malfunction, or drift in accuracy between calibration or periodic maintenance periods.

Fuel Burning Equipment

All types of fuel burning equipment have certain features in common: air registers or dampers, and parts subject to intense heat which are inaccessible during operation. These parts should be inspected annually for deformation, oxidation, or failure, or more frequently if outages occur. Hourly observation of fires is recommended to detect overheating of stoker, gas, oil or pulverized coal burner parts or unusual flame patterns which may indicate burner problems.

Gas Firing: Poor flame conditions may result from plugging of gas orifices in ring burners as a result of carbonization of light oil fractions contained in the gas which condenses when the gas is expanded and cooled in passing through orifices or pressure reducing stations. This condition is likely to develop in the lower burners of multiburner installations. Whenever burners are taken out of service in a multiburner boiler, the idle burner should be observed to determine that enough cooling air is supplied to prevent overheating of burner parts exposed to flame.

Oil Firing: Oil gun tips will require periodic removal for cleaning. The frequency is a function of oil type and atomization method. The frequency is best determined by plant experience and may vary from once per shift to monthly. Tips should be checked for wear, nicks, or cracking.

Oil pressure should be checked at individual burners to assure that reasonable oil distribution is being achieved. Atomizing steam or air pressure, if used, should conform to manufacturer's recommendation.

Stoker Firing: Mechanical condition of fuel distributors and grates should be checked if fires are burning unevenly. Excessive sifting through grates may indicate need for grate maintenance if fines content of coal is not abnormally high.

If an overfire air system is installed, the condition of nozzles should be determined during annual outages. During operation, the performance should be checked to determine that the required air flow is being developed.

Many stokers are equipped with cinder reinjection systems for returning ash collected in boiler hoppers or mechanical collectors to the furnace for reburning. This system can reduce unburned carbon heat losses by 2 to 5% and should be kept in good operating condition. Daily checks should be made to assure that hoppers are being emptied. Cold hopper spouts usually mean no flow. It is also usually possible to observe the flyash stream being injected into the furnace for flow indication. However, unless each hopper has a separate injection nozzle, there is no assurance that flyash is being reinjected from all hoppers.

Pulverized Coal Firing: The abrasive nature of pulverized coal requires some maintenance on deflectors or distributors which may be incorporated in the burner design. Oxidation may also occur.

These conditions should be checked during annual outages or more frequently if possible. Coke buildup on burner throats may occur with worn burner parts. Air register adjustment may be helpful in correcting coke formation conditions.

Heat Transfer Equipment

The principal efficiency-related maintenance aspects of heat transfer surfaces are surface cleanliness and good flue gas flow distribution. Flue gas-side cleanliness can be maintained by soot blowers and/or periodic water washing or mechanical cleaning. Water-side cleanliness can be maintained by proper water treatment and periodic chemical or mechanical cleaning.

Many older boiler designs used refractory or metal baffles to provide cross flow of gases in boiler tube banks. These baffles require periodic inspection and maintenance to provide the desired gas flow pattern. Missing baffle tile or plate will result in gas bypassing around heat transfer surfaces and increased boiler outlet temperature. More recent boiler designs are arranged for cross flow without requiring baffles. Gas distribution problems still can occur if localized heavy ash deposits plug gas flow lanes.

The boiler operator's log will indicate the rate of increase of exit gas temperature or draft loss which would determine the need for boiler cleaning. A sudden increase in boiler exit gas temperature possibly accompanied by a noticeable decrease in draft loss would indicate damage to cross flow baffles.

Many plants which have experienced problems with internal deposits leading to furnace tube failures have installed thermocouples to measure tube temperature at the failure location. If the tubes are clean, these thermocouples will read only a few degrees above saturation temperature. As internal deposits form, the thermocouple readings will become higher and indicate the need for chemical cleaning. If the annual inspection discloses significant internal deposits or there is evidence of tube wall overheating (bulges or swelling), furnace wall tube thermocouples would provide warning of poor water side conditions in the future.

Specific efficiency-related maintenance items to look for during an inspection outage are:

1) Soot or ash deposit formations on furnace walls which may indicate a need for burner adjustment or repair.
2) Evidence of furnace wall overheating (bulges or swelling of tubes).
3) General gas-side cleanliness of boiler, superheater, economizer and air heater surfaces,

and areas of heavy deposits which are not being cleaned by sootblowers.

4) Alignment of superheater elements (misalignment could contribute to ash plugging or increase gas flow resistance).
5) Condition of baffles or dampers in gas passages or air ducts.
6) Evidence of sootblower erosion on coal fired units which may be caused by high blowing pressure or excessive operation of blowers.
7) Water-side inspection to establish cleanliness of boiler and mechanical condition of drum internals.

Fans

Some forced draft fans with curved or air foil blades lose considerable capacity if the blades become dirty. This type of fan should be cleaned as required to maintain capacity. Inlet screens should be examined frequently for blockage by rags, paper, or other material which would restrict air entry.

Air and Gas Ducts

Leakage from air ducts will increase fan requirements and may change fuel/air ratio depending on the control system design. Inward leakage on balanced draft units will increase induced draft fan loading and may restrict unit capacity.

Insulation

Heat losses from uninsulated ducts, boiler surfaces, or steam lines can increase heat losses significantly in addition to creating personnel hazards and uncomfortable working conditions. Missing insulation should be replaced as required.

SUMMARY OF ROUTINE REQUIREMENTS AND RECORD KEEPING

As described previously a boiler efficiency improvement program must include two aspects: 1) action to bring the boiler to peak efficiency and 2) action to maintain the efficiency at the maximum level. This section deals with the requirements for maintaining maximum boiler efficiency. The analysis, technology and testing procedures described must be applied to achieve optimum efficiency in this latter effort. This then is a convenient summary or check list of those actions required to maintain optimum efficiency.

Many operations must be performed to maintain a boiler at or near its peak efficiency. These start with having a working knowledge of the components associated with the boiler, keeping records, etc. and end with cleaning heat transfer surfaces, adjusting the air-fuel ratio, etc. These requirements can be divided into the following:

- General Requirements,
- Daily Requirements,
- Weekly Requirements,
- Monthly Requirements, and
- Annual Requirements.

A description of each of these requirements with a representative example of a problem follows.

General Requirements

The general requirements are those which are usually associated with any system. They eliminate much of the need to consult others about the system. These requirements are listed in Table 8.1.

Example 1: If the boiler operator has a problem starting a boiler on a cold day after the plant has been shut down for several days, where should the operator go for help?

The answer to this question will probably not be found in the start-up procedures. Therefore, the operator will have to consult the boiler manual. The answer may not be in the boiler manual if the manufacturer has not anticipated this problem. In this case, the operator should consult a field serviceman.

TABLE 8.1: GENERAL REQUIREMENTS

ITEM	DESCRIPTION	COMMENT
1	Information about boiler and subsystems	This information can be found in manuals furnished by manufacturers a. boiler b. auxiliary equipment.

(continued)

TABLE 8.1: (continued)

ITEM	DESCRIPTION	COMMENT
2	Boiler Operating Procedure	Prepare and post detailed instructions for a. start-up, b. normal operation, and c. shut-down.
3	Parts List	Parts list of major items expected to be replaced a. name of part and b. number identifying part.
4	Maintenance Records	Maintenance Records a. date of maintenance b. description of maintenance c. person performing maintenance d. part(s) replaced, repaired, and/or cleaned.
5	Lubrication	Follow manufacturer's recommended procedures in lubricating all components.

Source: PB-265 713

Daily Requirements

The daily requirements are listed in Table 8.2.

Example 2: If an increase of 40°F in the exhaust gas temperature is observed, what should be done to determine the cause of this increase? The exhaust gas temperature means very little when considered without additional information such as firing position, exhaust gas composition, and combustion air temperature. If the increase in exhaust gas temperature is not caused by any of these factors, then fouled heat transfer surfaces should be considered.

TABLE 8.2: DAILY REQUIREMENTS

ITEM	DESCRIPTION	COMMENT
1	Check temperature of exhaust gases at two different firings.	Compare temperatures with tests performed after annual cleaning (see Table 8.5 for reference data).
2	Check Steam Pressure	Is variation in steam pressure as expected under different loads? Wet steam may be produced if the pressure drops too fast (caused by excessive loading on the boiler).
3	Check for Unstable Water Level	Common causes of unstable water level are: a. contaminates in boiler such as oil, excessive solids, excessive feedwater treatment, etc. b. overload on boiler c. malfunctions in equipment such as feedwater pump, water level control, etc.
4	Check Burner	a. Are controls functioning properly? b. Is burner clean? The burner may need cleaning several times daily if no. 6 fuel oil is used.
5	Check Motors and Auxiliary Equipment	Check to see that motors and auxiliary equipment are functioning properly.
6	Check Air Temperature in Boiler Room	The air temperature in the boiler room should not exceed or drop below recommended limits. These limits are determined by measuring the composition of the exhaust gases.
7	Check Blowdown	a. Manual Blowdown -- The frequency and amount of blowdown will depend upon the amount and condition of the feedwater. Check to see that blowdown valve does not leak. b. Automatic Blowdown -- Check operation of system and make sure that excessive blowdown does not occur.
8	Records	Keep daily records on a. Type and amount of fuel used b. Exhaust gas temperature and position of firing c. Boiler room temperature at time (b) is measured.

Source: PB-265 713

Weekly Requirements

Items listed under weekly requirements should be performed at a designated time each week. These requirements are listed in Table 8.3.

Example 3: If an operator measures, while burning No. 2 fuel oil, an oxygen content of 6% in the exhaust gases, how should this be corrected? The oxygen in the exhaust gases can be reduced to 2% by increasing the flow rate of fuel.

TABLE 8.3: WEEKLY REQUIREMENTS

ITEM	DESCRIPTION	COMMENT
1	Check Exhaust Gas Composition and Temperature	a. Measure exhaust gas composition and temperature at selected firing positions. b. Recommended percentages of oxygen and carbon dioxide in exhaust gases are: FUEL — O_2(%) — CO_2(%) Natural Gas — 1-1/2 — 10 No. 2 Fuel Oil — 2 — 11-1/2 No. 6 Fuel Oil — 2-1/2 — 12-1/2 Coal — 3-1/2 — 13-1/2 These percentages may vary due to variations in composition of fuel.
2	Check Relief Valve	a. Open valve manually while boiler is operating and observe steam pressure and force required. b. Check to see if relief valve leaks.
3	Check Water Level Control	Stop feedwater pump and allow control to stop fuel flow to burner. Do not allow water level in boiler to drop below the recommended low level.
4	Check Pilot and Burner Assemblies	a. Clean pilot and burner assemblies following recommended procedures. b. Check assemblies i. Spark gap ii. Condition of electrode iii. Condition of burner
5	Check Boiler Operating Characteristics	a. Stop fuel flow to burner and observe flame failure (characteristics and timing). b. Start boiler and observe characteristics of flame.

Source: PB-265 713

Monthly Requirements

The monthly requirements are listed in Table 8.4.

Example 4: What method is commonly used for removing dissolved oxygen from feedwater? The method is heating with steam.

TABLE 8.4: MONTHLY REQUIREMENTS

ITEM	DESCRIPTION	COMMENT
1	Check Blowdown and Water Treatment Procedures	a. Determine if blowdown procedure is adequate to prevent buildup of solids in boiler. b. Determine if water treatment procedure is adequate to prevent undesirable effects (see Chapter 6).
2	Exhaust Gases	a. Measure exhaust gas composition (O_2 and CO_2) and temperature over entire firing range. b. Compare composition and temperature readings with those of previous months and reference data (see Table 8.5)
3	Combustion Air Supply	a. Check combustion air inlet to boiler room to be sure adequate openings exist. b. Check combustion air inlet to boiler and clean if fouled.
4	Check Fuel System	Check pressure gages, pumps, filters, and transfer lines. Clean filters as required.
5	Check Belts and Packing Glands	a. Check belts for damage and proper tension. b. Check packing glands for proper compression and leakage.
6	Check for Air Leaks	a. Around access openings b. Around flame scanner assembly.

Source: PB-265 713

Annual Requirements

The annual requirements are listed in Table 8.5.

Example 5: Should the temperature of the exhaust gases be lower or higher after performing the annual requirements? The temperature should be lower because the heat transfer surfaces are clean.

TABLE 8.5: ANNUAL REQUIREMENTS

ITEMS	DESCRIPTION	COMMENT
1	Clean Waterside Surfaces	Follow manufacturer's recommended procedure in cleaning and preparing waterside surfaces.
2	Clean Fireside	Follow manufacturer's recommended procedure in cleaning and preparing fireside surfaces.
3	Repair Refractories on Fireside	Use recommended materials and procedure to repair refractory.
4	Relief Valve	Remove and recondition or replace.
5	Feedwater System	a. Clean condensate receivers, and deaeration system b. Clean and recondition feedwater pumps.
6	Fuel System	Clean and recondition system; pumps, filters, burner, pilot, oil preheaters, oil storage tanks, etc.
7	Electrical Systems	a. Clean all electrical terminals. b. Check electronic controls and replace any defected parts. c. Check mercury switches and replace if deterioration has occurred.
8	Hydraulic and Pneumatic Valves	Check operation and repair any leaks.
9	Start-up and Operation	Follow start-up and operation procedures.
10	Exhaust Gases	a. Make adjustments to give desired exhaust gas composition. b. Record composition, firing position and temperature (Reference Data).

Source: PB-265 713

REFERENCES

(1) Cato, G.A., et al, *Field Testing: Application of Combustion Modifications to Control Pollutant Emissions from Industrial Boilers - Phase II,* KVB, Inc. Report to U.S. Environmental Protection Agency, No. EPA 600/2-76-086a, April 1976.

(2) Thompson, R.E., et al, *A Study to Assess the Potential for Energy Conservation Through Improved Industrial Boiler Efficiency,* Report to Federal Energy Administration, Contract No. C-04-50085-00, Oct. 1976.

Determination of Combustion Efficiency

The material in this chapter was excerpted from a report by Engineering Extension Service, Auburn University (PB-265 713) and a Federal Power Commission report (TID-27600).

TESTING BOILER PERFORMANCE

The performance of a boiler may be determined by two methods. The direct method measures steam generator efficiency. To perform this test, measurements must be made to determine the input and output flow rates of water and fuel as well as temperatures, pressures, and fuel composition. The indirect method measures combustion efficiency. To perform this test it is necessary to obtain a fuel analysis and an exhaust gas analysis.

Testing Procedure for the Direct Method

The boiler should be tested at different flow rates of steam over the range in which the boiler is operated. The flow rates between successive tests should vary about 10% of rated capacity.

The following testing procedure is recommended:

1) install calibrated instruments (see Figure 9.1 for an illus-

tration of installation which does not require shutdown, and Figure 9.2).

2) operate boiler for one hour at desired load prior to starting test,

3) mark water level in boiler,

4) perform test for approximately one hour, reading

   temperatures every 10 minutes,
   pressures every 10 minutes,
   flow rates every 15 minutes or total flow during the one hour test,

5) stop test when water level in boiler is in the same position as start of test.

The flow rates, temperatures, and pressures should remain essentially constant during the test. This is necessary to reduce the uncertainties in evaluating the results of the test.

Testing Procedure for the Indirect Method

As with the direct method, the boiler should be tested at several load conditions over the entire range of operation when applying the indirect method.

The following testing procedure is recommended:

1) Install calibrated instruments (see Figure 9.3),
2) Operate the boiler at specified load for approximately one hour prior to making the test,
3) Make an Orsat analysis of the exhaust gases,
4) Take a fuel sample (should be collected according to procedures recommended by fuel analysis laboratory) and measure fuel temperature and chemical composition,
5) Record data
   a. Intake air temperature, pressure, and relative humidity,
   b. Flue gas temperature and exhaust gas composition,
   c. Fuel chemical composition.

The flow rates and temperatures should remain essentially constant during the test.

FIGURE 9.1: BYPASS LOOP CONTAINING FLOW METER, PRESSURE, AND TEMPERATURE INSTRUMENTS

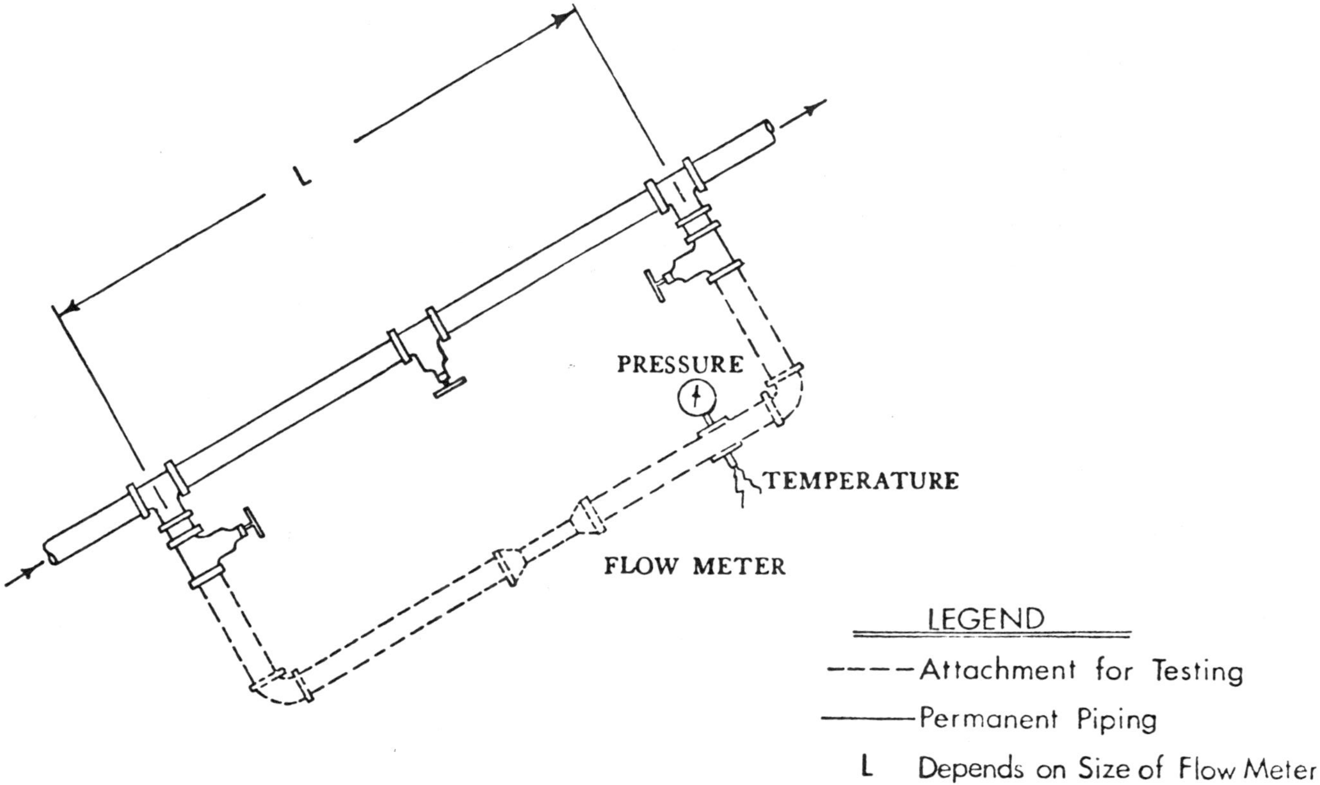

Source: PB-265 713

FIGURE 9.2: INSTRUMENTS TO CALCULATE EFFICIENCY OF BOILER BY DIRECT METHOD

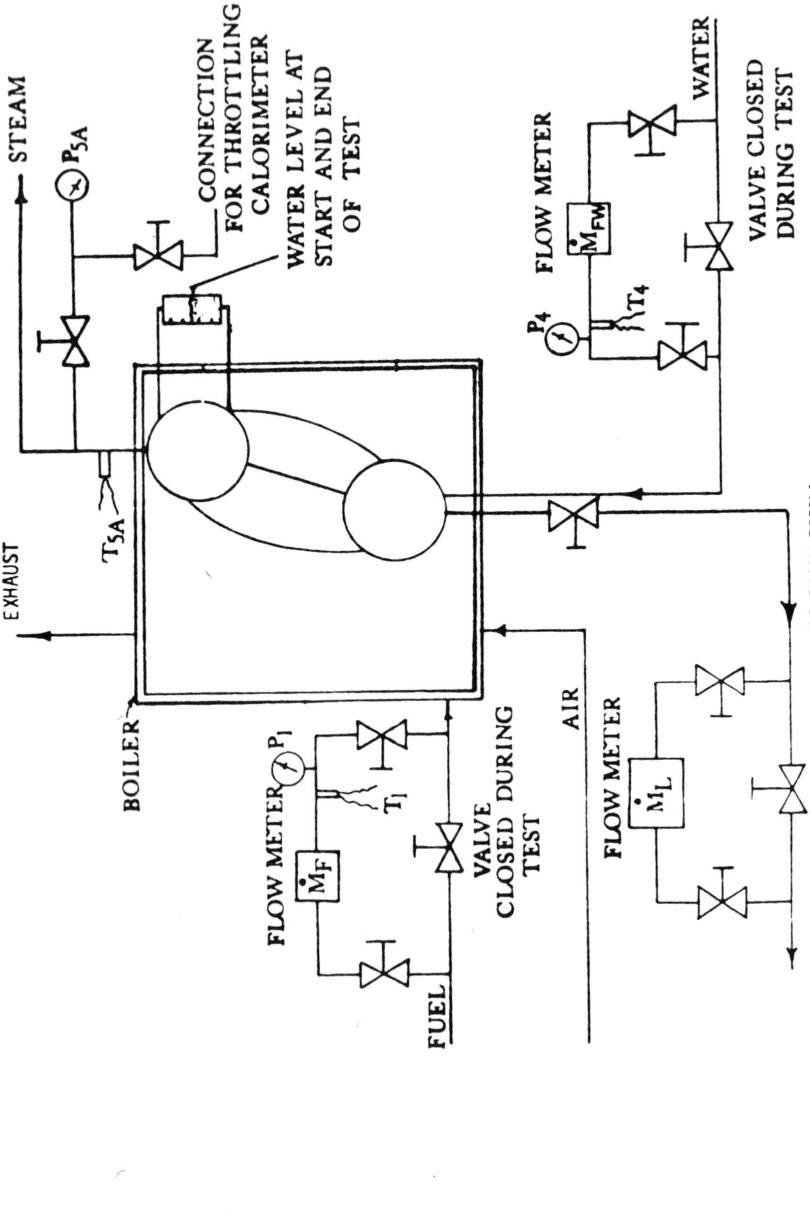

Source: PB-265 713

FIGURE 9.3: INSTRUMENTS TO CALCULATE COMBUSTION EFFICIENCY FOR INDIRECT METHOD

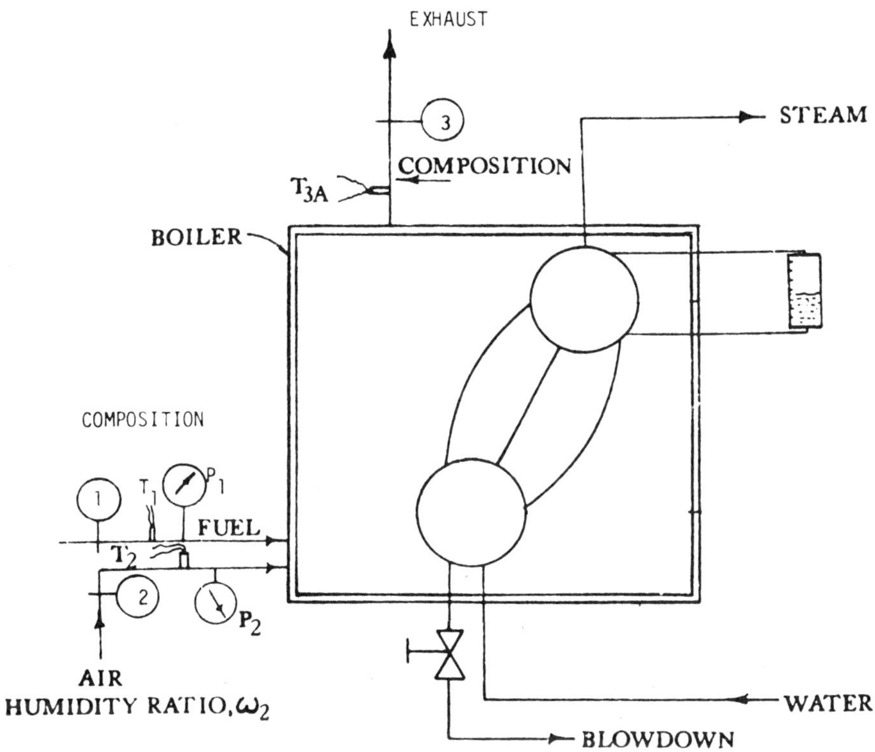

Source: PB-265 713

MEASUREMENT METHODS, DEVICES AND TECHNIQUES

This section is concerned with methods of determining the air/fuel ratios of a process and adjusting for optimum performance. For a discussion of methods of calculating combustion efficiency, please see the following chapter.

Methods of Determining Air/Fuel Ratio

The three basic ways of determining the A/F ratio of the combustion process in industrial heating equipment are

- Metering the air and fuel flow at each individual burner.

- Analyzing the flue products of the entire combustion system at the flue(s).
- Visually observing the flame characteristics.

Because of variations in equipment, processes, and controls, none of these methods will be universally applicable to all adjustments. As will be shown, however, the most typical determination of combustion ratio will be made by a combination of these three methods.

Metering air and fuel flows at each burner may appear to be the most desirable. In doing so the required A/F ratio can be set at each burner of a multiburner system. Metering has some drawbacks, however. For example, adjustment upstream of individual burners does not account for leakage of air into the furnace through doors and openings. Thus, after all burners have been set at stoichiometric ratios, considerable oxygen may still exist in the flue products as a result of air infiltration into the furnace.

Also, this technique requires individual air and fuel meters and valves at each burner, and these are not always included as an integral part of the burner. The addition of these meters may prove quite expensive and difficult. Finally, this technique assumes complete combustion by the burner. Owing to improper burner design or maintenance, it could be possible to supply a stoichiometric ratio to the burner only to have incomplete combustion products generated by the burner.

A variation of the individual burner metering technique would be the metering of air and fuel that is supplied to an entire zone of burners. Several burners are quite often manifolded together and supplied with air and fuel from common lines. Although air- and fuel-metering devices are sometimes included in the zone supply lines, the other drawbacks to metering still exist.

The most economical and reliable method of determining the combustion ratio of an entire system in industrial situations is to analyze the flue products. This technique involves the quantitative measurement of one or more of the products of combustion. This measurement can then be used to determine the A/F ratio of the entire combustion process. Effects of air infiltration or improper burner performance are, under these conditions, accounted for. This could be considered a drawback, since the exact cause of off-ratio combustion, e.g., air leakage, a bad burner, or improper A/F supplies, is not pinpointed.

Techniques for adjustment, which will be explained later, however, will serve to minimize these errors by reverting to other methods for additional adjustment and control. Flue-products analysis, however, will be the primary method of determining the need for A/F ratio adjustment.

Admittedly, visual observation of flame characteristics is not an accurate method of determining combustion conditions; however, such observations can serve to expedite adjustments. In cases of atmospheric burner equipment, visual observation is even necessary to establish a starting point for adjustment.

Flow Measurement

If air- and fuel-metering devices are available in a burner system, they can be used to adjust the combustion ratio. The most common types of meters found in combustion applications are obstruction meters, rotameters, and quantity meters. Three common forms of obstruction meters are the venturi, the flow nozzle, and the flat-plate orifice (see Figure 9.4).

In each case the basic meter acts as an obstacle placed in the path of the flowing fluid and thus causes localized changes in velocity. Concurrent with velocity change, there will be a pressure change. A certain portion of the pressure drop becomes unrecoverable; therefore the output pressure will always be less than the input pressure. Pressure taps are provided on the meter assembly. The pressure drop across the meter is related to the flow through the meter by the general equation

$$Q = K(\Delta P)^{1/2}$$

where Q is the flow rate through the orifice (generally expressed in cubic feet per hour or per minute) and K is a constant defined by the meter characteristics, the type of fluid, and the units of flow and pressure. If meters are supplied with the burner, a flow chart for that particular meter should also be provided. The flow chart is simply a graph of the equation $Q = K(\Delta P)^{1/2}$ on log-log graph paper and would appear as shown in Figure 9.5.

This flow chart shows the flow through the meter plotted against the pressure drop (ΔP) across the meter. Pressure drop is normally expressed in terms of inches of water column and is therefore measured by a water manometer.

Rotameters are most often used to meter liquid flows but can also be

FIGURE 9.4: TYPICAL OBSTRUCTION METERS

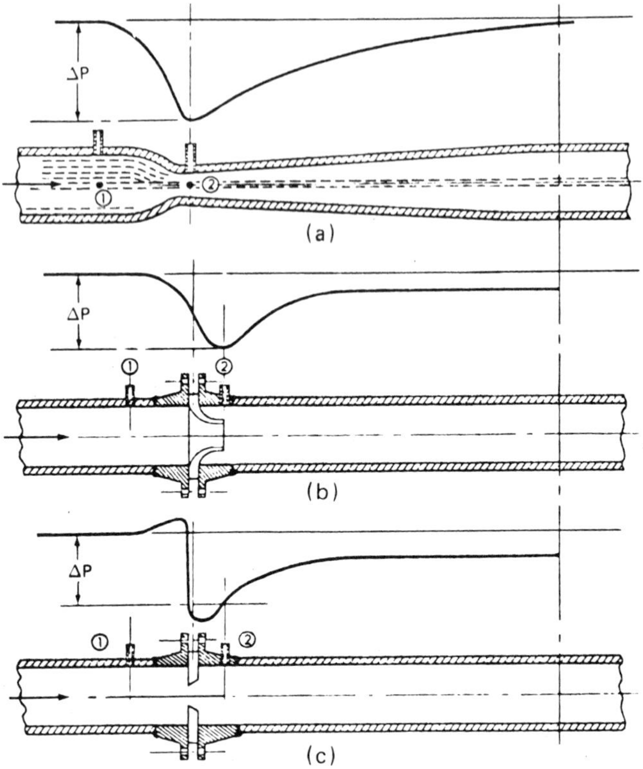

(a) Venturi. (b) Flow nozzle. (c) Flat-plate orifice.

Source: TID-27600

FIGURE 9.5: TYPICAL ORIFICE FLOW CHART

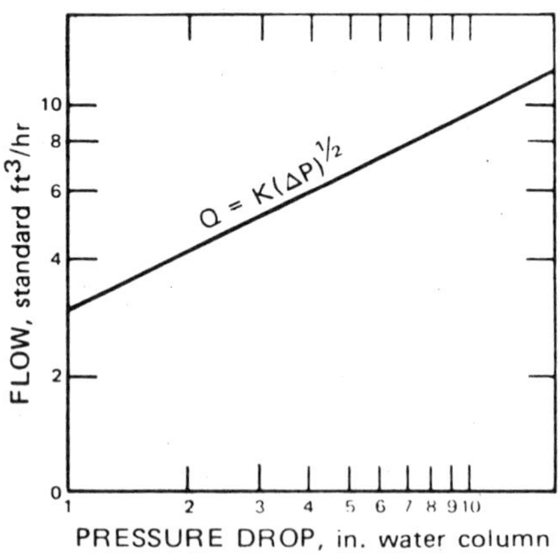

Source: TID-27600

used for gas measurements. The essential parts of the rotameter are the float and tapered tube in which the float is free to move. It is designed so that the forces acting on the float are proportional to the upward fluid flow through the meter. Hence its position will determine the flow rate. The tube is normally calibrated in specific units of flow for a specific fluid. Some meters measure only accumulated volume. The volume must be measured over a specific period of time to determine flow rates. Since rotameters are not continuous metering devices, their use in measuring flow rates of combustion systems is restricted.

Flue-Product Measurements

Since flue-product analysis will be the primary method of determining combustion ratio, it is important to know and understand the significance of combustion products. As described earlier, perfect combustion will produce flue products containing carbon dioxide (CO_2), nitrogen (N_2), and water vapor (H_2O). In lean combustion free oxygen (O_2) will also exist in the flue products. In rich combustion combustibles in the form of hydrogen (H_2) and carbon monoxide (CO) will appear along with carbon dioxide, nitrogen, and water vapor.

Figure 9.6 shows the dry flue-products analysis of natural gas and air at various A/F ratios. Figure 9.7 shows the same analysis for propane. Water vapor (H_2O) is not shown in these figures because most of the water vapor in the combustion products will condense when the gases are cooled down to room temperature for analysis.

Note that for both fuels shown CO, H_2, N_2, and CO_2 appear at substoichiometric ratios. Only CO_2 and N_2 exist at stoichiometric conditions, and CO_2, N_2 and O_2 appear at ratios above stoichiometric conditions. Also, only CO_2 and N_2 appear over the whole range of ratios, i.e., both above and below the stoichiometric ratio.

Properly adjusted burner equipment should operate with a slight amount of excess air in the flue gases. This amount should be preferably between 2.5 and 5% and should normally not exceed 10%. Accordingly, the flue gases will contain between 0.5 and 1% oxygen and should not exceed 2%.

Measurement of the oxygen percentage in flue products will prove to be the simplest and most applicable single-constituent measurement. First, the change in percent oxygen vs A/F ratio close to the stoichiometric ratio is significant. Second, low oxygen percentage indicates near stoichiometric combustion regardless of the type of fuel. Measuring the percent of oxygen in flue products also ensures that adjustments are being made in relatively safe combustion conditions where no combustibles are present. The absence of oxygen in flue products does not, of course, signify stoichiometric conditions.

Thus, when one uses oxygen percentage as a determination of A/F ratio and no oxygen can be measured, the first adjustment to the combustion system should be toward lean combustion. These adjustments would continue until a small amount of oxygen is detected, which indicates near but above stoichiometric combustion.

FIGURE 9.6: FLUE-PRODUCT ANALYSIS OF NATURAL GAS vs AIR/FUEL RATIO

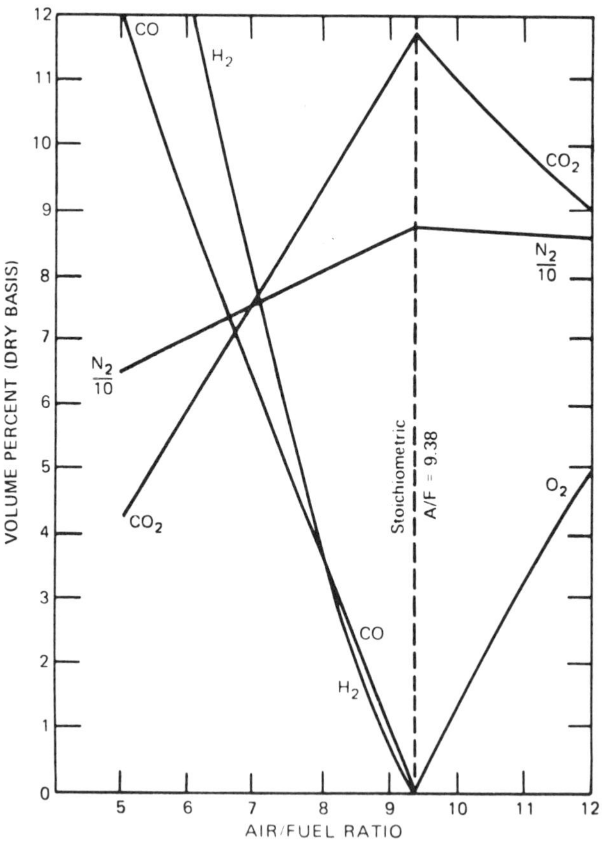

Source: TID-27600

FIGURE 9.7: FLUE-PRODUCT ANALYSIS OF PROPANE vs AIR/FUEL RATIO

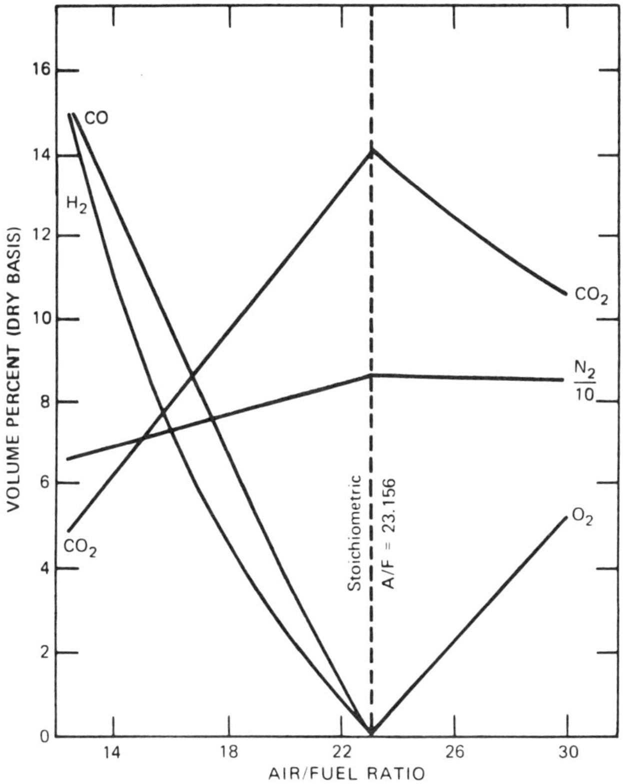

Source: TID-27600

The ideal situation would be to measure at least two constituents of combustion products. This would allow immediate and exact indications of A/F ratios. However, so that general adjustment instructions can be applied with a minimum of measuring equipment, the measurement systems described in Table 9.1 are suggested.

The adjustments to specific burner equipment, outlined in later chapters of this manual, are based on oxygen determination. If additional measuring equipment is available or if the particular instrument used is capable of measuring more than one constituent, the additional measurements should indeed be taken and used as a check in the final determination.

TABLE 9.1: MEASUREMENT SYSTEMS

Measurement system	Advantages	Disadvantages
Measure oxygen only	Most applicable because only one instrument is needed and low oxygen indicates proper A/F ratio.	Substoichiometric conditions are not defined.
Measure oxygen and total combustibles	Most preferable because there is an instrument that will measure both oxygen and combustibles. This defines total combustion range.	
Measure oxygen and carbon dioxide	Defines entire combustion range.	Two instruments required.
Measure oxygen and carbon monoxide	Defines entire combustion range.	Two instruments required.
Measure carbon dioxide only	Only one instrument required.	Particular information on the fuel and equilibrium flue-gas products should define whether combustion is above or below stoichiometric condition.

Source: TID-27600

There are times, e.g., in low-temperature combustion such as with immersion-heating and atmosphere-drying ovens, when the products of combustion may contain both oxygen and combustibles, and thus measuring only oxygen would give no indication of improper combustion. It would be advisable, therefore, to measure combustibles or CO_2 contents of flue products and refer to flue-product analysis information to ensure proper combustion.

Devices Used for Measurements

As is also described in the next chapter, an untold number of instruments for measuring individual products of combustion are available.

They range from the simple test-tube detector to the complex gas chromatograph system. The cost, too, may range from a few dollars to many thousands. This section briefly describes a few of the devices that are felt to be the most suitable for measuring flue-product constituents in everyday industrial-type environments.

The devices listed in Table 9.2 are considered most applicable for the measurements required in A/F adjustments.

TABLE 9.2: MEASURING DEVICES

Device	Measures	Remarks
Orsat analyzer	CO_2, O_2, CO, H_2, and CH_4 (N_2 by diff.)	Selective absorption of individual constituents by chemical reaction.
Heat prover	O_2, percent combustible	Measures oxygen by paramagnetic principle or electrochemical reaction. Measures combustibles by catalytic combustion of the sample with air.
Oxygen analyzer	O_2	Measures oxygen by paramagnetic or electrochemical principles.
Infrared analyzer	CO, CO_2, or CH_4	Measures the amount of absorption of infrared radiation by a particular gas, absorption being proportional to the gas concentration.
Gas absorption	O_2 or CO_2	Works on same principle as Orsat except that only one gas constituent, either CO_2 or O_2, is measured by a given unit.

Source: TID-27600

An Orsat consists of a series of five glass tubes containing chemicals that selectively absorb CO, CO_2, and O_2. Hydrogen and methane are determined by controlled combustion of remaining combustibles with a measured amount of oxygen and subsequent absorption. Sampling for Orsat analysis is noncontinuous, i.e., a measured volume is drawn from the flue products and completely analyzed before another sample can be taken. Operation is completely manual, and accuracy depends on careful measurement of gas volumes throughout the procedure. A well-trained operator, using sound operating techniques, can obtain very repetitive and accurate gas analyses. Maintenance is generally negligible, but chemicals must be periodically replaced, and an oxygen source is required for H_2 and CH_4 determination. A three-tube Orsat, which can measure CO_2 and O_2 only, is often used. Of the five devices considered, the Orsat is of intermediate cost.

The heat prover consists of two individual measuring systems. One system measures free oxygen in the sample with meter readout on a dual scale normally ranging from 0 to 5% and 5 to 21% oxygen. The other system measures the total combustibles in the sample with meter readout on dual scales from 0 to 3% and 3 to 15% combustibles.

Calibration procedures are quite simple but should be repeated periodically. Normal calibration procedures do not require any reference gases. Sampling is continuous, which proves quite beneficial in observing immediate variations in flue-product analysis resulting from adjustments. Accuracy on a well-maintained heat prover is good. Of the five devices, the heat prover is considered intermediate in cost.

Oxygen analyzers measure the total amount of oxygen in the sample gas. Ranges may vary from 0 to 1% to 0 to 100%, with dual ranges normally standard. Standard air may serve for calibration of instruments having a 0 to 25% range. Operation and calibration procedures are quite simple. Sampling is continuous. Accuracy on a well-maintained unit is excellent. Compared to the other devices discussed, the cost of an oxygen analyzer is intermediate to high.

Infrared analyzers are designed to measure CO_2, CO, or CH_4. One instrument can only measure one gas. The instruments are normally single range but can be supplied in a wide variation of ranges serving the typical flue-product percentage. Accuracy on a well-maintained unit is excellent. Calibration requires a zero gas (nitrogen) and a span gas, which should be near full-scale concentration. Calibration and operating procedures are quite simple. Sampling is continuous. Readout is normally by meter. Compared to the other devices, the cost of infrared instruments is high.

As with the Orsat, gas-absorption units measure CO_2 or O_2 content by absorption of the gas by a liquid. A given unit measures only one constituent, either CO_2 or O_2. Operation is manual and extremely simple, requiring little technique or experience. Maintenance is negligible. Chemicals must be replaced periodically. Sampling is noncontinuous.

Compared to other devices discussed, the cost of absorption units is low.

Measurement Techniques

The manufacturers' operating instructions will clearly define the instruments and operating procedures required for accurate measurements. Careful consideration must also be given to flue-product sampling techniques to ensure measurements representative of the combustion process. In most cases these techniques will involve preliminary measurement, and in some cases adjustment, of furnace pressure. Although the procedure for measuring furnace pressure is quite universal, adjustment of furnace pressure is very dependent on process, furnace design, and type of burner equipment. This discussion is concerned only with measuring furnace pressure.

Generally, furnace pressure will have to be set at a very low positive or negative value (0.1 inch H_2O or less) and measurements made at a specific location in the furnace (normally at hearth level). Because of the low pressures involved, the most applicable and economical measuring device will be the inclined manometer, which is available in many sizes and pressure ranges. The accuracy of the inclined manometer should be within the specified furnace pressure. Because of the high temperatures involved, a length of heat-resistant alloy tubing long enough to extend from the specified point of pressure measurement to 4 or 5 ft out into ambient temperature (see Figure 9.8) will be required.

FIGURE 9.8: SCHEMATIC OF FURNACE PRESSURE MEASUREMENT

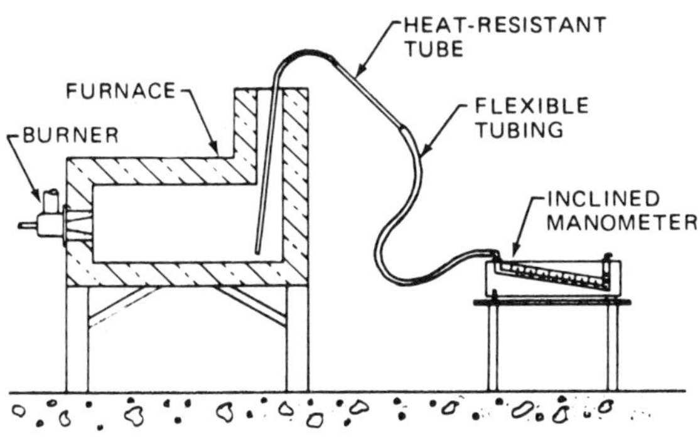

Source: TID-27600

From this point standard rubber tubing can be used to connect the tube to the manometer inlet port. The inclined manometer should be properly leveled and zeroed according to operating instructions before the pressure impulse is connected. Periodic checks on zeroing are recommended because of the low-pressure measurements required.

Once furnace pressure has been set by adjusting furnace or burner conditions, sampling of flue products may begin. In addition to the measuring equipment described earlier, a length of heat-resistant alloy tubing, a water-cooled probe, or a ceramic probe and sufficient rubber tubing will be necessary to conduct the flue-product sample from the furnace to the instrument. Care should be taken to ensure leakproof connections between the probe, rubber tubing, and instrument.

Of the five instruments discussed, only heat provers, oxygen analyzers, and infrared analyzers require a continuous sample. A sample pump is usually an integral part of the instrument. If a pump is not included, one should be supplied. Continuous-sampling instruments will also require a rather specific sample throughput to maintain accuracy. Again, if a flow indicator is not an integral part of the instrument, one must be supplied. A rotameter-type meter of approximately 4 to 6 ft^3/hr range will usually suffice. A filter in the sample line is also necessary to prevent dirt, carbon, and moisture from building up in the instrument.

A typical hookup for a continuous-sampling instrument is shown in Figure 9.9, where sample pump, filter, and flowmeter are all considered to be external to the instrument. The arrangement of the various items shown is considered optimum since only the lines from the probe to pump inlet are under a suction. Although all leakage should be avoided, leakage of air into the sampling line immediately invalidates the measurement. Leakage occurring on the pressure side does not affect the results unless the amount of sample reaching the instrument is inadequate.

Sampling of flue products with Orsats and absorption units is not a continuous process. Extreme care should be taken to purge the sampling line adequately by drawing and discharging numerous samples before retaining a sample for analysis. Generally, pumps, filters, and flowmeters are not required with these instruments. Samples are pulled through the instruments by a pumping action inherent in the system's operation. Use of a pump and filter, as shown in Figure 9.6, is possible, however, and can considerably reduce the sampling time.

FIGURE 9.9: SCHEMATIC OF FLUE-PRODUCT SAMPLING SYSTEM

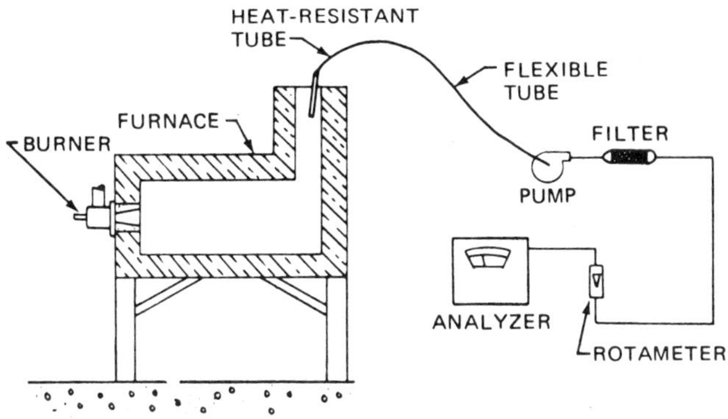

Source: TID-27600

Probe location is very important in flue-product sampling. Care must be taken to ensure that the gas sample obtained is representative of actual conditions. The best location for the probe is in a flue opening or stack, or at some point in a furnace where gases are well mixed. The probe should extend far enough in to avoid drawing in ambient air. It should not be placed near a burner or above an opening. Placing it too near a burner may result in pulling a sample exclusively from that particular burner. Placing it above or near an opening or door could result in obtaining oxygen concentrations that are really higher than the actual products in the stack.

A final consideration in obtaining representative flue-product samples is furnace operating input. Because of conditions related to furnace pressure, the flue-product sample should be obtained with the combustion system at high fire and at or near normal operating temperatures. The relationship between furnace pressure and burner input will be described in later chapters of this manual. Regardless of specific pressure requirements, however, the most representative flue-product sample will be obtained at full burner input.

In practical situations burners may not be operating at full input long enough for reliable gas samples to be taken. If this is the case, the furnace temperature should be lowered 50° to 100°F below normal set point. After the temperature has stabilized at the lower set point, an increase to the normal set point should allow sufficient time at high fire to obtain good samples. This procedure may need to be repeated after each adjustment.

Lowering the furnace temperature may adversely affect the heating process; thus a process engineer should be consulted before it is done. Adjustments should be made with no work in the furnace.

Calculation of Boiler Efficiency

The material in this chapter was excerpted from a report by Engineering Extension Service, Auburn University (PB-265 713).

DIRECT METHOD OF TESTING

Determination of Boiler Efficiency

The efficiency of a boiler by the direct method is defined as

[1] $$n \equiv \frac{\dot{M}_S\,(h_{5A} - h_4)}{\dot{M}_F\,(HHV)}$$

where

HHV = higher heating value of fuel (Btu/lb)
h_4 = enthalpy of water entering boiler (Btu/lb)
h_{5A} = enthalpy of steam leaving boiler (Btu/lb)
$\dot{M}_F$ = mass rate of flow of fuel (lb/h)
$\dot{M}_S$ = mass rate of flow of steam (lb/h);
($\dot{M}_S = \dot{M}_{FW}$ - Blowdown), and
n = efficiency of boiler.

The instruments needed to calculate the efficiency are shown in Figure 8.2.

With no blowdown, the measurements needed are:

For Water Entering Boiler

Flow rate, $\dot{M}_{FW}$, $\dot{M}_{FW} = \dot{M}_S$

Temperature (T_4)

Pressure (P_4)

For Steam Leaving Boiler

Temperature (T_{5A})

Pressure (P_{5A}) or Quality (X_{5A}) if steam is saturated

For Fuel

Flow rate, M_F

Temperature (T_1)

Pressure (P_1)

Higher Heating Value of Fuel (HHV)

The higher heating value (HHV) of the fuel is usually obtained by sending a sample to a fuel analysis laboratory. The fuel sample should be collected according to recommended procedures.

The efficiency can be calculated using equation [1] and the measurements shown above.

Summary of Measurements

The measurements necessary to determine the efficiency by the direct method are given in Table 10.1.

TABLE 10.1: MEASUREMENTS FOR DIRECT METHOD

FEEDWATER

FLOW RATE	TEMP. (°F)	PRESS. (lb/in^2g)

(continued)

TABLE 10.1: (continued)

STEAM

FLOW RATE	TEMP. (°F)	PRESS. (lb/in^2g)	QUALITY (if saturated)

FUEL

FLOW RATE	TEMP. (°F)	PRESS. (lb/in^2g)	HHV $\left(\frac{Btu}{lb}\right)$

Source: PB-265 713

Accuracy of Instrumentation

Involved in any calculations, based on the measurement of different quantities, are the errors associated with the measuring equipment. Errors encountered in the measurement of data tend to compound each other. Thus, even though the instruments used to obtain the data are fairly accurate, large cumulative errors may result due to the compounding effect of the individual measurements. It is therefore important to make each data reading as accurate as possible to reduce the inaccuracy in the desired calculation.

Accuracy of Calculations and Effects

The efficiency of a boiler is given by equation [1]. It is seen from this equation that four data measurements ($\dot{M}_S$, Δh_W, $\dot{M}_F$, and HHV) and their associated errors formulate an expression describing the boiler efficiency.

The relationship between the percent error associated with any one of the four determining factors and the boiler efficiency is presented in Figure 10.1. Note that the x-axis is the percent error associated with only one quantity ($\dot{M}_S$, $\dot{M}_F$, Δh_W or HHV) with the other three assumed 100% accurate. For example, with a ±2% accuracy associated with the HHV data (all other quantities assumed 100% accurate) for an 80% efficient boiler, one would expect an error of ±1.6% in the actual calculated boiler efficiency. As the accuracies of the three remaining quantities are determined, they are combined according to the following equation:

$$\text{rms percent error} = \sqrt{(\text{error in } \dot{M}_S)^2 + (\text{error in } \dot{M}_F)^2 + (\text{error in } \Delta h_W)^2 + (\text{error HHV})^2}$$

Example: If associated with the ±2% HHV data an error of ±1.5% is anticipated in $\dot{M}_F$, ±1.5% in $\dot{M}_S$ and ±1% in Δh_W, the total error from the readings would be:

$$\text{rms error} = \sqrt{(1.5)^2 + (1.5)^2 + (1.0)^2 + (2.0)^2} = 2.92\% \text{ error.}$$

Entering Figure 10.1 with this value and an 80% boiler efficiency, a ±2.34% error is expected in the actual boiler efficiency.

This is clearly an undesirably large spread from 77.66 to 82.34% and indicates the need for careful attention to the associated errors used in determining the boiler efficiency.

FIGURE 10.1: EFFECT OF ERROR IN MEASURING $\dot{M}_S$, $\dot{M}_F$, Δh, OR HHV ON BOILER EFFICIENCY

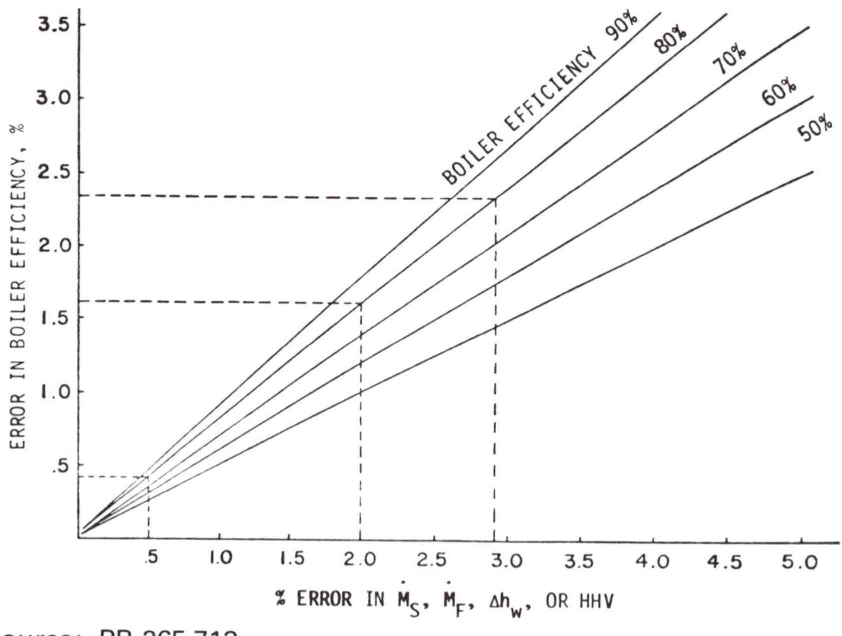

Source: PB-265 713

Individual Errors

Figure 10.1 was presented to show the significant effects that each of the component data inaccuracies impose on the boiler efficiency. The accuracies associated with the various measuring devices are considered below.

Mass Flow Rates: The accuracy of instruments measuring mass flow are usually rated as some percentage of the actual flow or a percentage of full scale flow. Those meters having an error based on the actual flow rate are usually more accurate than those having a fixed error independent of the actual flow rate. Thus, with an instrument indicating an accuracy of ±½% of actual flow rate one can enter Figure 10.1 directly and predict, for an 80% boiler efficiency, an error of ±0.4% in efficiency.

Figure 10.2 can be used to determine the percent error associated with the measurement of the flow rate with the second type of flow meter.

FIGURE 10.2: ERROR ASSOCIATED WITH MEASURING FLOW RATE WITH FLOW METER OF SECOND TYPE

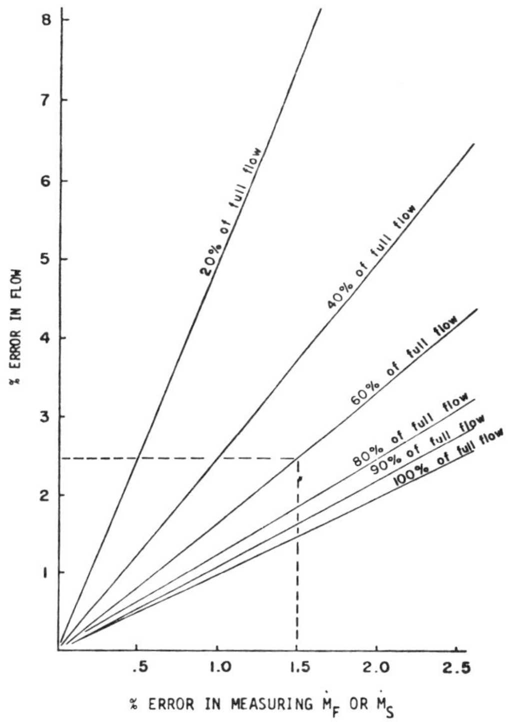

Source: PB-265 713

The various lines indicate the actual flow rate as a percent of a full flow capacity.

Example—Consider a meter accurate to ±1.5% of full scale flow. M_F or $\dot{M}_S$ is measured by this meter and only 60% of the total flow measuring capacity is utilized. Therefore the expected error in flow rate would be obtained by entering Figure 10.2 with a percent error in flow of ±1.5%, intersecting the line for 60% of full flow and reading across to a value of ±2.5% error in flow. This error would then be used to enter Figure 10.1 and for a boiler efficiency of 80% one would expect a resulting error of ±2% in actual boiler efficiency. If the other error terms are to be included, they should be combined and the rms percent error used to enter Figure 10.1.

Change in Enthalpy (Δh_W): The enthalpy of the incoming water and outgoing steam is a function of the temperature and pressure at each state. With these two parameters known, the enthalpy of the liquid or steam may be obtained from Steam Tables.

The enthalpy of the water entering the boiler is assumed to be that of a saturated liquid at the state temperature, so pressure is not a determining factor. The error associated with measuring the temperature of the water entering the boiler usually causes small errors in the enthalpy.

The enthalpy of the steam leaving the boiler is about 20 times larger than the enthalpy of the water entering. At the steam exiting state, however, both temperature and pressure are needed to determine the enthalpy, if the steam is superheated. The errors in Δh_W versus the percent error in pressure and temperature measurements are indicated in Figure 10.3. This figure clearly indicates that the usual accuracies in measuring pressure and temperature have a minor effect on Δh_W.

Example—Consider a temperature reading of 80°F ±1.5% and a pressure of 120 psia ±1% on the inflow water and 10°F superheated steam (350°F) ±1.0% at 120 psia ±1.5% exiting the boiler. The inflow conditions are assumed accurate as the errors effect a small enthalpy term thus producing small errors. State 1 is the enthalpy at T = 80°F. The errors associated with state 2 are found in Figure 10.3. For the temperature error, entering with ±1.0% one notes an associated error in Δh_W ±0.19%. Entering with the ±1.5% error in pressure yields a ±0.097% error in Δh_W. Thus, the combined error in Δh_W is still very small.

Higher Heating Value (HHV): Samples of the fuel are analyzed in a laboratory for higher heating value. By this means, values may be ob-

FIGURE 10.3: EFFECT OF ERRORS IN MEASURING TEMPERATURE AND PRESSURE ON ENTHALPY OF STEAM

a.

b.

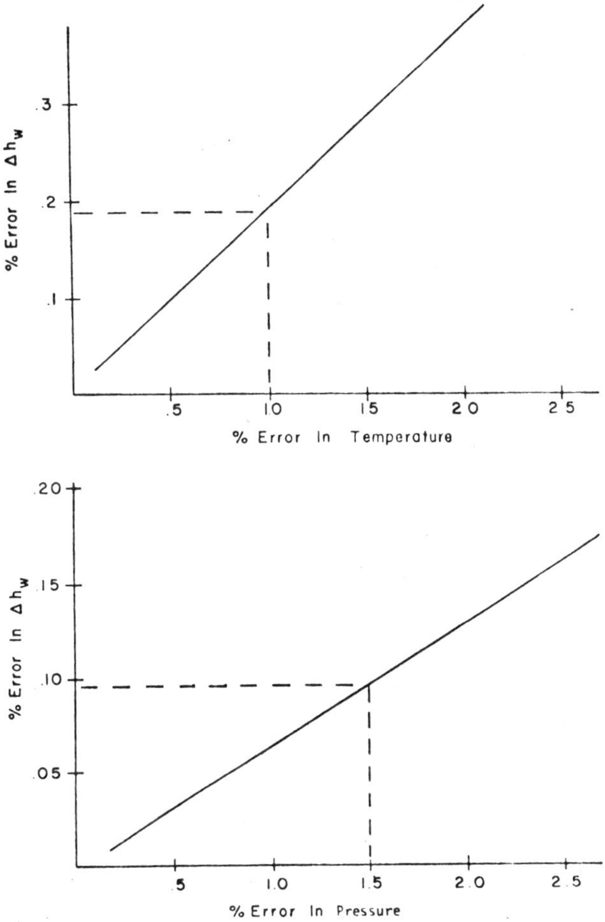

Source: PB-265 713

tained to an accuracy specified by the testing laboratory. Typical HHVs cited for various fuels by fuel suppliers are usually of insufficient accuracy to allow the boiler efficiency to be calculated with the necessary precision.

Types of Instruments

Table 10.2 shows instruments for analyzing boilers.

TABLE 10.2: INSTRUMENTS FOR DIRECT METHOD

INSTRUMENT	DESCRIPTION	MAX. ALLOW ERROR	USE
Flow Meter	Turbine Type	$\pm$ 0.25% of flowrate	Measure fuel feedwater flow rate
Thermocouple (T.C.)	Type K Chrom-Alum	$\pm$ 4°F	Measure temp. of fuel feedwater flow rate
Thermocouple Readout	Milivolt Meter	$\pm$ 0.25%	Reads mv. output of T.C.
Pressure Gauge	Hiess Gauge	$\pm$ 1/2% of full scale	Measure pressure fuel feedwater

Source:PB-265 713

INDIRECT METHOD OF TESTING

Determination of Boiler Combustion Efficiency

Combustion efficiency is defined as the energy released in combustion divided by the higher heating value of the fuel. Mathematically, it is represented as

[2] $$n_c = \frac{|h_{products} - h_{reactants}|}{HHV}$$

where,

$| \; |$ = indicates the absolute value of the function

$h_{products}$ = is the enthalpy of the products of combustion per pound of fuel

$h_{reactants}$ = is the enthalpy of the reactants per pound of fuel

HHV = is the higher heating of the fuel.

Summary of Measurements

The data required to calculate the combustion efficiency are given in Table 10.3.

TABLE 10.3: MEASUREMENTS FOR INDIRECT METHOD

COMBUSTION AIR

TEMP. (°F)	PRESS. (lb/in²a)	HUMIDITY (%)

EXHAUST GASES

COMPOSITION (MOL PERCENT)				TEMP. (°F)
CO_2	CO	O_2	OTHERS	

FUEL PROPERTIES

COMPOSITION (MOL PERCENT)								HHV $(\frac{Btu}{lb})$	TEMP. (°F)	EQUIVALENT HYDROCARBON COMPOSITION	
α*	N_2	H_2O	O_2	C	H_2	S_2	CO_2			$C_A H_B$	
										A	B

* Mols of Equivalent Hydrocarbon

Source: PB-265 713

The measurements needed are:

For Intake Air

Temperature, T_2

Relative humidity, ω_2

Pressure, P_2

For Exhaust Gases

Temperature, T_{3A}
Composition

For Fuel

Temperature, T_1
Composition
Higher heating value
Pressure, if gaseous.

The higher heating value and the composition of a fuel are best obtained by sending a sample to a fuel analysis laboratory. The fuel sample should be collected according to recommended procedures.

The relative humidity of the combustion air can be determined by measuring the dry bulb and wet bulb temperatures of the air in the boiler room. The dry bulb temperature is read by conventional methods. The wet bulb temperature is read with a sling psychrometer or by placing a wet (with water) piece of cotton around the bulb of a conventional thermometer and blowing air over the cotton. These two temperatures can be used to obtain the humidity ratio from a psychrometric chart.

The combustion efficiency can be calculated using equation [2] and the measurements shown above.

Error Analysis

The variables considered in the error analysis are CO_2, CO, O_2, and flue gas temperature. These variables were chosen because they are the most significant variables in the combustion efficiency calculation.

The method of analysis is essentially the same as that for the direct method. The expected error in efficiency is taken to be the root-mean-square of the sum of the individual errors.

Figure 10.4 shows the relative size of the errors contributed by only one measurement. The only significant composition variable is the CO_2 since it is the largest single measurement and contributes significantly to the absolute enthalpy of the products of combustion.

FIGURE 10.4: ERROR IN CALCULATED EFFICIENCY DUE TO INACCURACY IN MEASUREMENT OF ONE SPECIES IN EXHAUST GAS

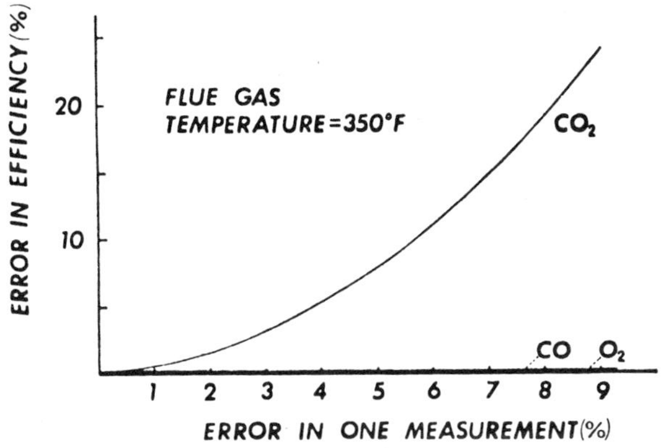

Source: PB-265 713

Figure 10.5 shows the combined effect of all three composition variables. As expected the effects of the error closely follow the single effect of the CO_2 error.

FIGURE 10.5: ERROR IN CALCULATED EFFICIENCY DUE TO INACCURACY IN MEASUREMENT OF ALL SPECIES IN EXHAUST GAS

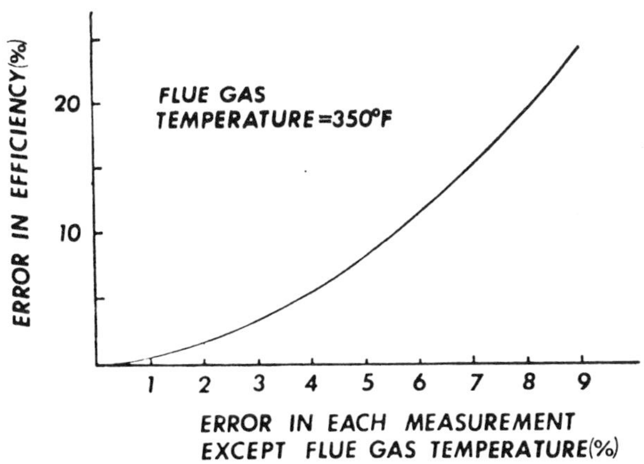

Source: PB-265 713

Figure 10.6 shows the effect of an error in flue gas temperature. This effect is pronounced because the absolute enthalpy of the products is a function of temperature only.

FIGURE 10.6: ERROR IN CALCULATED EFFICIENCY DUE TO INACCURACY IN FLUE GAS TEMPERATURE MEASUREMENT

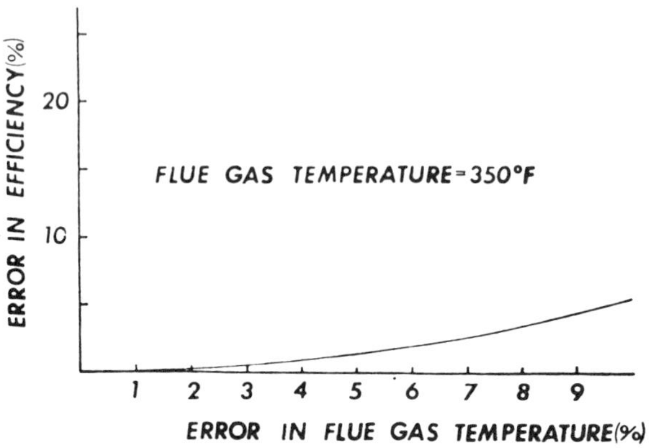

Source: PB-265 713

Figure 10.7 shows the effect of all the variables on accuracy.

FIGURE 10.7: ERROR IN CALCULATED EFFICIENCY DUE TO INACCURACY IN MEASUREMENT OF FLUE GAS TEMPERATURE AND SPECIES IN EXHAUST GAS

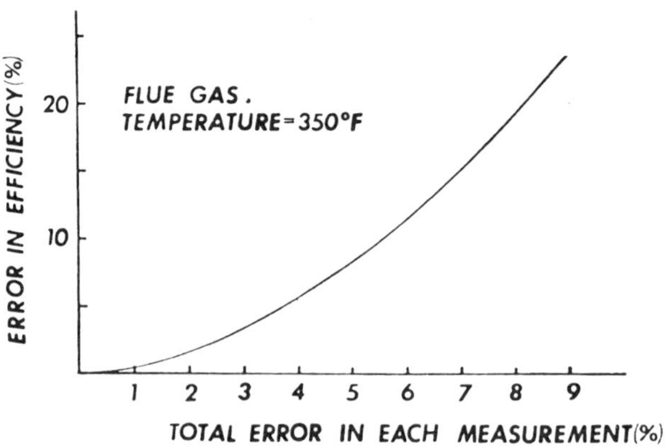

Source: PB-265 713

In estimating the error it should be noted that, as a general trend the error in efficiency increases as flue gas temperature increases. Thus curves for errors at higher flue gas temperatures would show greater errors in efficiency.

Types of Instruments

Table 10.4 shows instruments which can be used to determine boiler combustion efficiency.

TABLE 10.4: INSTRUMENTS FOR DIRECT METHOD

INSTRUMENT	DESCRIPTION	MAX. ALLOW ERROR	USE
Flue Gas Analyzer	Orsat Analyzer	± 1% of actual volume	Measure CO_2, CO, & O_2 in flue gas
Oxygen Analyzer	Hagan Cell	± 4% of actual volume	Measure O_2 in flue gas

Source: PB-265 713

CHARTS FOR DETERMINING EFFICIENCY OF FUELS

Considerable detail is required in order to calculate combustion efficiency. In many situations it is only necessary to be able to determine the combustion efficiency when the amounts of combustibles and CO in the flue gas are negligible. Under this assumption the efficiency is a function only of the percentage of oxygen in the dry flue products and the flue gas temperature for a given fuel.

The following tables (10.5 to 10.7) can be used with the above assumption to determine combustion efficiency for a typical natural gas, #2 and #6 fuel oil, respectively. Note that each table gives data grouped under flue gas temperature as a function of either excess air or percentage oxygen in the dry products.

TABLE 10.5: COMBUSTION EFFICIENCY TABLE FOR NATURAL GAS

TFLUE = 250.00

% EXCESS AIR	% EXCESS OXYGEN	EFFICIENCY (%)
0.0	0.0	83.061
4.48	1.00	82.933
9.44	2.00	82.792
14.94	3.00	82.635
21.09	4.00	82.460
28.02	5.00	82.264
35.86	6.00	82.041
44.82	7.00	81.786
55.16	8.00	81.492
67.23	9.00	81.148
81.48	10.00	80.743
98.58	11.00	80.256
119.48	12.00	79.662
145.60	13.00	78.919
179.18	14.00	77.964
223.93	15.00	76.691

TFLUE = 300.00

% EXCESS AIR	% EXCESS OXYGEN	EFFICIENCY (%)
0.0	0.0	82.089
4.48	1.00	81.925
9.44	2.00	81.743
14.94	3.00	81.541
21.09	4.00	81.315
28.02	5.00	81.061
35.86	6.00	80.773
44.82	7.00	80.444
55.16	8.00	80.065
67.23	9.00	79.622
81.48	10.00	79.099
98.58	11.00	78.472
119.48	12.00	77.705
145.60	13.00	76.746
179.18	14.00	75.514
223.93	15.00	73.872

(continued)

TABLE 10.5: (continued)

TFLUE = 350.00

% EXCESS AIR	% EXCESS OXYGEN	EFFICIENCY (%)
0.0	0.0	81.110
4.48	1.00	80.908
9.44	2.00	80.685
14.94	3.00	80.437
21.09	4.00	80.160
28.02	5.00	79.849
35.86	6.00	79.496
44.82	7.00	79.093
55.16	8.00	78.628
67.23	9.00	78.085
81.48	10.00	77.443
98.58	11.00	76.674
119.48	12.00	75.733
145.60	13.00	74.558
179.18	14.00	73.047
223.93	15.00	71.034

TFLUE = 400.00

% EXCESS AIR	% EXCESS OXYGEN	EFFICIENCY (%)
0.0	0.0	80.121
4.48	1.00	79.882
9.44	2.00	79.617
14.94	3.00	79.324
21.09	4.00	78.995
28.02	5.00	78.626
35.86	6.00	78.208
44.82	7.00	77.729
55.16	8.00	77.178
67.23	9.00	76.534
81.48	10.00	75.773
98.58	11.00	74.861
119.48	12.00	73.746
145.60	13.00	72.352
179.18	14.00	70.561
223.93	15.00	68.173

(continued)

TABLE 10.5 (continued)

TFLUE = 450.00

% EXCESS AIR	% EXCESS OXYGEN	EFFICIENCY (%)
0.0	0.0	79.123
4.48	1.00	78.846
9.44	2.00	78.539
14.94	3.00	78.199
21.09	4.00	77.819
28.02	5.00	77.392
35.86	6.00	76.907
44.82	7.00	76.353
55.16	8.00	75.714
67.23	9.00	74.969
81.48	10.00	74.088
98.58	11.00	73.032
119.48	12.00	71.740
145.60	13.00	70.127
179.18	14.00	68.052
223.93	15.00	65.287

TFLUE = 500.00

% EXCESS AIR	% EXCESS OXYGEN	EFFICIENCY (%)
0.0	0.0	78.114
4.48	1.00	77.799
9.44	2.00	77.451
14.94	3.00	77.064
21.09	4.00	76.632
28.02	5.00	76.145
35.86	6.00	75.594
44.82	7.00	74.964
55.16	8.00	74.237
67.23	9.00	73.390
81.48	10.00	72.388
98.58	11.00	71.186
119.48	12.00	69.717
145.60	13.00	67.881
179.18	14.00	65.522
223.93	15.00	62.377

(continued)

TABLE 10.5: (continued)

TFLUE = 550.00

% EXCESS AIR	% EXCESS OXYGEN	EFFICIENCY (%)
0.0	0.0	77.096
4.48	1.00	76.742
9.44	2.00	76.351
14.94	3.00	75.917
21.09	4.00	75.432
28.02	5.00	74.887
35.86	6.00	74.268
44.82	7.00	73.562
55.16	8.00	72.746
67.23	9.00	71.795
81.48	10.00	70.671
98.58	11.00	69.323
119.48	12.00	67.675
145.60	13.00	65.616
179.18	14.00	62.969
223.93	15.00	59.440

Source: PB-265 713

TABLE 10.6: COMBUSTION EFFICIENCY TABLE FOR NO. 2 FUEL OIL

TFLUE = 250.00

% EXCESS AIR	% EXCESS OXYGEN	EFFICIENCY (%)
0.0	0.0	90.501
4.68	1.00	90.359
9.86	2.00	90.202
15.61	3.00	90.028
22.04	4.00	89.833
29.27	5.00	89.614
37.46	6.00	89.366
46.83	7.00	89.082
57.63	8.00	88.755
70.24	9.00	88.373
85.13	10.00	87.922
103.00	11.00	87.380
124.84	12.00	86.718
152.13	13.00	85.891
187.20	14.00	84.828
233.96	15.00	83.412

(continued)

TABLE 10.6: (continued)

TFLUE = 300.00

% EXCESS AIR	% EXCESS OXYGEN	EFFICIENCY (%)
0.0	0.0	89.503
4.68	1.00	89.319
9.86	2.00	89.117
15.61	3.00	88.892
22.04	4.00	88.641
29.27	5.00	88.358
37.46	6.00	88.038
46.83	7.00	87.672
57.63	8.00	87.249
70.24	9.00	86.757
85.13	10.00	86.175
103.00	11.00	85.476
124.84	12.00	84.622
152.13	13.00	83.555
187.20	14.00	82.184
233.96	15.00	80.356

TFLUE = 350.00

% EXCESS AIR	% EXCESS OXYGEN	EFFICIENCY (%)
0.0	0.0	88.495
4.68	1.00	88.270
9.86	2.00	88.022
15.61	3.00	87.746
22.04	4.00	87.438
29.27	5.00	87.092
37.46	6.00	86.699
46.83	7.00	86.250
57.63	8.00	85.732
70.24	9.00	85.128
85.13	10.00	84.414
103.00	11.00	83.557
124.84	12.00	82.510
152.13	13.00	81.202
187.20	14.00	79.521
233.96	15.00	77.279

(continued)

TABLE 10.6: (continued)

TFLUE = 400.00

% EXCESS AIR	% EXCESS OXYGEN	EFFICIENCY (%)
0.0	0.0	87.477
4.68	1.00	87.210
9.86	2.00	86.916
15.61	3.00	86.589
22.04	4.00	86.224
29.27	5.00	85.813
37.46	6.00	85.347
46.83	7.00	84.815
57.63	8.00	84.201
70.24	9.00	83.484
85.13	10.00	82.637
103.00	11.00	81.622
124.84	12.00	80.380
152.13	13.00	78.829
187.20	14.00	76.835
233.96	15.00	74.177

TFLUE = 450.00

% EXCESS AIR	% EXCESS OXYGEN	EFFICIENCY (%)
0.0	0.0	86.448
4.68	1.00	86.140
9.86	2.00	85.799
15.61	3.00	85.421
22.04	4.00	84.997
29.27	5.00	84.522
37.46	6.00	83.982
46.83	7.00	83.366
57.63	8.00	82.655
70.24	9.00	81.825
85.13	10.00	80.845
103.00	11.00	79.668
124.84	12.00	78.231
152.13	13.00	76.435
187.20	14.00	74.126
233.96	15.00	71.048

(continued)

TABLE 10.6: (continued)

TFLUE = 500.00

% EXCESS AIR	% EXCESS OXYGEN	EFFICIENCY (%)
0.0	0.0	85.409
4.68	1.00	85.058
9.86	2.00	84.671
15.61	3.00	84.240
22.04	4.00	83.759
29.27	5.00	83.217
37.46	6.00	82.604
46.83	7.00	81.903
57.63	8.00	81.094
70.24	9.00	80.150
85.13	10.00	79.035
103.00	11.00	77.697
124.84	12.00	76.062
152.13	13.00	74.019
187.20	14.00	71.392
233.96	15.00	67.891

TFLUE = 550.00

% EXCESS AIR	% EXCESS OXYGEN	EFFICIENCY (%)
0.0	0.0	84.359
4.68	1.00	83.965
9.86	2.00	83.531
15.61	3.00	83.047
22.04	4.00	82.508
29.27	5.00	81.900
37.46	6.00	81.212
46.83	7.00	80.425
57.63	8.00	79.518
70.24	9.00	78.459
85.13	10.00	77.208
103.00	11.00	75.707
124.84	12.00	73.873
152.13	13.00	71.581
187.20	14.00	68.634
233.96	15.00	64.707

Source: PB-265 713

TABLE 10.7: COMBUSTION EFFICIENCY TABLE FOR NO. 6 FUEL OIL

TFLUE = 250.00

% EXCESS AIR	% EXCESS OXYGEN	EFFICIENCY (%)
0.0	0.0	91.835
4.73	1.00	91.690
9.96	2.00	91.528
15.77	3.00	91.349
22.27	4.00	91.149
29.58	5.00	90.923
37.86	6.00	90.668
47.32	7.00	90.376
58.23	8.00	90.040
70.97	9.00	89.647
86.02	10.00	89.183
104.07	11.00	88.626
126.14	12.00	87.946
153.71	13.00	87.095
189.16	14.00	86.002
236.40	15.00	84.545

TFLUE = 300.00

% EXCESS AIR	% EXCESS OXYGEN	EFFICIENCY (%)
0.0	0.0	90.827
4.73	1.00	90.639
9.96	2.00	90.430
15.77	3.00	90.199
22.27	4.00	89.941
29.58	5.00	89.650
37.86	6.00	89.321
47.32	7.00	88.944
58.23	8.00	88.510
70.97	9.00	88.003
86.02	10.00	87.405
104.07	11.00	86.686
126.14	12.00	85.809
153.71	13.00	84.712
189.16	14.00	83.301
236.40	15.00	81.422

(continued)

TABLE 10.7: (continued)

TFLUE = 350.00

% EXCESS AIR	% EXCESS OXYGEN	EFFICIENCY (%)
0.0	0.0	89.809
4.73	1.00	89.578
9.96	2.00	89.323
15.77	3.00	89.039
22.27	4.00	88.722
29.58	5.00	88.366
37.86	6.00	87.962
47.32	7.00	87.500
58.23	8.00	86.968
70.97	9.00	86.347
86.02	10.00	85.613
104.07	11.00	84.732
126.14	12.00	83.655
153.71	13.00	82.310
189.16	14.00	80.581
236.40	15.00	78.276

TFLUE = 400.00

% EXCESS AIR	% EXCESS OXYGEN	EFFICIENCY (%)
0.0	0.0	88.780
4.73	1.00	88.507
9.96	2.00	88.204
15.77	3.00	87.868
22.27	4.00	87.492
29.58	5.00	87.069
37.86	6.00	86.590
47.32	7.00	86.043
58.23	8.00	85.412
70.97	9.00	84.675
86.02	10.00	83.804
104.07	11.00	82.760
126.14	12.00	81.483
153.71	13.00	79.888
189.16	14.00	77.838
236.40	15.00	75.105

(continued)

TABLE 10.7: (continued)

TFLUE = 450.00

% EXCESS AIR	% EXCESS OXYGEN	EFFICIENCY (%)
0.0	0.0	87.741
4.73	1.00	87.424
9.96	2.00	87.074
15.77	3.00	86.684
22.27	4.00	86.249
29.58	5.00	85.760
37.86	6.00	85.205
47.32	7.00	84.572
58.23	8.00	83.840
70.97	9.00	82.987
86.02	10.00	81.979
104.07	11.00	80.770
126.14	12.00	79.292
153.71	13.00	77.444
189.16	14.00	75.070
236.40	15.00	71.906

TFLUE = 500.00

% EXCESS AIR	% EXCESS OXYGEN	EFFICIENCY (%)
0.0	0.0	86.691
4.73	1.00	86.330
9.96	2.00	85.932
15.77	3.00	85.489
22.27	4.00	84.994
29.58	5.00	84.437
37.86	6.00	83.806
47.32	7.00	83.085
58.23	8.00	82.254
70.97	9.00	81.283
86.02	10.00	80.137
104.07	11.00	78.761
126.14	12.00	77.080
153.71	13.00	74.978
189.16	14.00	72.278
236.40	15.00	68.678

(continued)

TABLE 10.7: (continued)

TFLUE = 550.00

% EXCESS AIR	% EXCESS OXYGEN	EFFICIENCY (%)
0.0	0.0	85.630
4.73	1.00	85.225
9.96	2.00	84.778
15.77	3.00	84.281
22.27	4.00	83.726
29.58	5.00	83.101
37.86	6.00	82.394
47.32	7.00	81.585
58.23	8.00	80.652
70.97	9.00	79.563
86.02	10.00	78.277
104.07	11.00	76.733
126.14	12.00	74.847
153.71	13.00	72.490
189.16	14.00	69.460
236.40	15.00	65.422

Source: PB-265 713

Economics Evaluation

The material in this chapter was excerpted from a report by Engineering Extension Service, Auburn University (PB-265 713).

ENERGY CONSERVATION ECONOMICS

This chapter describes the basic technique for the economic evaluation of any given energy conservation opportunity. Not only will these methods allow the boiler operator to determine whether a given project or procedure should be undertaken but it will also allow him to effectively communicate with managers whose support is essential if an effective energy conservation program is to be achieved.

Definitions

FC is First Cost. This item is the initial cost of any capital expenditure which is made. For example, if an automatic blowdown control system is to be installed, FC would be the labor and material charge required to install this equipment.

AOC is Annual Operating Cost. The AOC includes all costs associated with maintaining and operating the installed system. For example, with the automatic blowdown system this charge would include service labor and materials to maintain the control system. This item must be estimated. Assistance in estimating can be obtained from the equipment supplier.

AFS is Annual Fuel Savings. This item is simply the amount of fuel which is saved by installation of the device per year. In order to establish a true value for this item, measurement of performance of the existing system and either measured or theoretically calculated performance of the installed system must be made. A comparison of these performance values allows the fuel savings to be determined. In most cases guess work is not satisfactory for determination of this item. Taking the installation of automatic blowdown control, one would compare the performance in terms of mass loss with existing system and that of the controlled system as supplied by the equipment manufacturer. The energy saved would be the difference between the enthalpy of blowdown water and feedwater multiplied by the mass reduction of blowdown. The AFS would be the energy saved per year divided by the boiler efficiency times the higher heating value (HHV) of the fuel.

PFP is Projected Fuel Price. The PFP is the estimated average fuel price over the lifetime of the investment. Admittedly, it is very difficult to assess. It almost certainly will be conservative to estimate the fuel cost over the next 5-10 years at say 150% of today's cost.

EL is Estimated Life. EL is the length of time that the investment is expected to be in use.

S is Net Annual Savings. This item is the annual savings considering operating cost and is given by: (AFS $\times$ PFP) $-$ AOC.

DC is Depreciation Charge. DC is the annual depreciation of the investment. For the blowdown control system example, straight line depreciation would be FC/EL.

Types of Economic Measures

It is clear that money saved today is more valuable than at a later date. This concept, however, is not to be confused with inflation. A simple way to understand this point is to compare the case of saving $100 now and putting it in a savings account at 6% interest versus the case of saving $100 next year and putting it in a savings account at 6% interest. At the end of any succeeding year, the difference between the accrued return on the money invested earlier and that invested later increases. Economic performance measures are classified according to whether account is not or is taken by the "time value of money" by first and second order of measures, respectively. It is emphasized that first order measures are used as rough guides to economic performance, but major expenditures justify use of more complex second order measures when first order economic performance is marginal.

In the following sections discussions of these types of measures will be given.

First Order Measures of Economic Performance

The most popular measure of this type is the Payback Period (PP) which is defined by

$$PP = \frac{FC}{S}$$

This measure allows one to determine if an investment can be recouped in a reasonable time compared to the estimated life (EL). Usually if the payback period (PP) is less than one-half of the estimated life (EL) the investment is considered potentially profitable. This measure does not allow comparison of investment opportunities with different estimated life (EL). Account of different estimated life (EL) can be taken through the Depreciation Charge (DC). Another first order measure is defined as the Return on Investment (ROI) which does allow different estimated life (EL) to be considered. It is defined by

$$ROI = \frac{S - DC}{FC} \times 100\%$$

The rate of return should be of at least the order of the best alternate investment opportunity (typically 15-20%).

As noted above these measures are to be used as simple indicators for screening possible energy conservation opportunities but second level measures may be required on major tasks.

Second Order Measures of Economic Performance

A reasonable value for the time value of money is usually specified as a discount rate (D). The discount rate (D) is taken to be the best internal return rate on any investment alternatives. Thus, a dollar earned in the future is discounted. There are several second level measures which utilize this concept. Only the most popular one is presented, namely the Benefit/Cost Ratio (B/C). In this measure a Present Worth Factor (PWF) is defined such that Present Value (PV) of all future savings is calculated as follows:

$$PV = PWF \times S$$

The present worth factor represents the number of years over which the net annual savings can be applied to arrive at the discounted value of these savings. The PWF is a function of both the discount rate (D) and the estimated life (EL). Table 11.1 (1) is provided to summarize this functional relationship.

TABLE 11.1: PRESENT WORTH FACTORS (PWF)

Lifetime (EL)	Discount Rate (D)				
	5%	10%	15%	20%	25%
1	0.952	0.909	0.870	0.833	0.800
2	1.859	1.736	1.626	1.528	1.440
3	2.723	2.487	2.283	2.106	1.952
4	3.546	3.170	2.855	2.589	2.362
5	4.329	3.791	3.352	2.991	2.689
6	5.076	4.355	3.784	3.326	2.951
7	5.786	4.868	4.160	3.605	3.161
8	6.463	5.335	4.487	3.837	3.329
9	7.108	5.759	4.772	4.031	3.463
10	7.722	6.145	5.019	4.192	3.571
11	8.306	6.495	5.234	4.327	3.656
12	8.863	6.814	5.421	4.439	3.725
13	9.394	7.103	5.583	4.533	3.780
14	9.899	7.367	5.724	4.611	3.824
15	10.380	7.606	5.847	4.675	3.859
16	10.838	7.824	5.954	4.730	3.887
17	11.274	8.022	6.047	4.775	3.910
18	11.690	8.201	6.128	4.812	3.928
19	12.085	8.365	6.198	4.843	3.942
20	12.462	8.514	6.259	4.870	3.954
21	12.821	8.649	6.312	4.891	3.963
22	13.163	8.772	6.359	4.909	3.970
23	13.489	8.883	6.399	4.925	3.976
24	13.799	8.985	6.434	4.937	3.981
25	14.094	9.077	6.464	4.948	3.985

Source: PB-265 713

The benefit/cost ratio (B/C) is then given by

$$B/C = \frac{PV}{FC} = \frac{PWF \times S}{FC}$$

Calculation for Energy Conservation Assessment

Example: *Basic Problem Specification*—A food processing plant includes a boiler with a rating of 8.82×10^6 Btu/hr used to produce 165 psia steam. The boiler is fired with #2 fuel oil and has a stack temperature of 600°F. There is a requirement for 70 gpm of hot water at 175°F which is currently generated in a hot water heater. The measured flue gas flow rate is 10,500 lbm/hr.

An energy conservation opportunity exists to recover energy from the flue gas in a heat exchanger by dropping the flue gas temperature to 220°F and heating the required water from 60° to 180°F. It is required to analyze the economics associated with this opportunity.

Flue gas has a specific heat of about 0.25 Btu/lb °F. Therefore, the amount of energy available in the flue gas is

$$(\dot{m}C_p\Delta T)_G = 10{,}500\ \frac{\text{lb}}{\text{hr}}\ 0.25\ \frac{\text{Btu}}{\text{lb }^\circ\text{F}}\ (600\text{-}220)^\circ\text{F} = 997{,}500\ \frac{\text{Btu}}{\text{hr}}$$

For continuous operation this would result in an Annual Fuel Savings of

$$\text{AFS} = 365\text{ days } 24\ \frac{\text{hours}}{\text{day}}\ 997{,}500\ \frac{\text{Btu}}{\text{hr}} = 8{,}738\text{ million Btu}$$

The amount of water that can be heated can be calculated by an energy balance on the boiler, namely

$$\dot{m}_W C_{pW}\,(\Delta T)_W = \dot{m}_G\ C_{pG}\ \Delta T_G$$

or

$$\dot{m}_W = \frac{997{,}500\text{ Btu/hr}}{1\ \dfrac{\text{Btu}}{\text{lb }^\circ\text{F}}\ 120^\circ\text{F}}$$

$$\dot{m}_W = 8{,}312\ \frac{\text{lb}}{\text{hr}} = 16.6\ \frac{\text{gal}}{\text{min}}$$

Therefore all the water can be used.

Discussions with a shell and tube heat exchanger manufacturer reveal that the heat exchanger required for this application has an installed first cost (FC) = $5,000. The annual operating charge (AOC) is about 10% of the First Cost (FC) so that AOC = $500. This cost includes maintenance and fan power cost. The heat exchanger manufacturer also gives the estimated life (EL) = 15 years. The appropriate discount rate (D) for the company is D = 20%. The projected fuel price is $1.50/million Btu.

Summary of Economic Data—The data measured and determined above can be summarized as follows:

FC = \$5,000 from manufacturer
AOC = \$375 from manufacturer
AFS = 8,730 million Btu
PFP = \$1.50/million Btu Best estimate
S = AFS x PFP - AOC
= 8730 x \$1.50 - \$375
= \$12,720 per year
D = 20%

First Order Measure of Performance—Payback Period (PP):

$$PP - \frac{FC}{S} = \frac{\$5,000}{12,720} = 0.39 \text{ year}$$

This is extremely good and indicates the project should definitely be undertaken.

Return on Investment (ROI):

$$DC = \frac{FC}{EL} = \frac{\$5,000}{15} = \$333$$

(The DC is based on straight line depreciation)

$$ROI = \frac{S - DC}{FC} \times 100$$

$$= \frac{12,720 - 333}{5,000} \times 100$$

$$= 247\% \text{ per year!}$$

This is well above any company's best internal rate of return and again indicates that this project should be undertaken.

Second Order Measure of Performance—From Table 11.1 for a discount rate of 20% and expected life of 15 years, the present worth factor (PWF) = 4.675 then,

$$PV = S \times PWF$$

$$= \$12{,}720 \times 4.675$$

$$= \$59{,}466$$

$$B/C = \frac{PV}{FC} = \frac{59{,}466}{5{,}000} = 11.89$$

This overwhelmingly indicates the attractiveness of this energy conservation opportunity. Any time the B/C ratio is greater than 1, the project is feasible.

REFERENCE

(1) Gatts, R., Massey, R. and Robertson, J.C., "Financial Evaluation Procedures," *Energy Conservation Program Guide for Industry and Commerce,* National Bureau of Standards, 1974, pp. 5.1 through 5.5.

Sources Utilized

The following reports were used in the preparation of this book:

PB-262 577

Industrial Boiler Users' Manual—Volume II. KVB, Inc. FEA Contract C-04-50085-00, Report No. FEA/D-77/026, January 1977.

PB-264 543

Guidelines for Industrial Boiler Performance Improvement (Boiler Adjustment Procedures to Minimize Air Pollution and to Achieve Efficient Use of Fuel) by M.W. McElroy and D.E. Shore, KVB, Inc. EPA Contract 68-02-1074, Report No. EPA-600/8-77-003a, January 1977.

PB-265 713

Measuring and Improving the Efficiency of Boilers—A Manual for Determining Energy Conservation in Steam Generating Power Plants, by Engineering Extension Service, Auburn University, Alabama. FEA Contract FEA-CO-04-50100-00, Report No. FEA/D-77/132, November 1976.

TID-27600

Maintenance and Adjustment Manual for Natural Gas and No. 2 Fuel Oil Burners. Bureau of Natural Gas, Federal Power Commission, and U.S. Department of Energy, 1977.

BOILER FUEL ADDITIVES FOR POLLUTION REDUCTION AND ENERGY SAVING 1978

Edited by R.C. Eliot

Energy Technology Review No. 33
Pollution Technology Review No. 53
Chemical Technology Review No. 120

Problems of incomplete combustion, soot deposition, and sludge formation are common to a greater or lesser degree with all hydrocarbon fuel oils, especially those sold for use in heavy duty installations, e.g., furnaces and boilers in power plants, large office buildings, hospitals, etc. The problem of particulate emissions is particularly acute, as these carbonaceous contaminants are currently a violation of local anti-pollution laws in many communities.

This book reflects the findings of investigations conducted for the U.S. Environmental Protection Agency and the Federal Energy Administration with the purpose of determining the effectiveness of chemical additives not only in reducing smoke and particulate emissions, but also as a means of increasing overall boiler efficiency, and thus conserving fuel. A survey of the recent patent literature is included.

Various kinds of compounds have shown merit as combustion improvers, including surfactants, organometallics, and low-molecular-weight polymers. Alkaline-earth and transition metal additives in concentrations of 20 ppm to 50 ppm of metal in the fuel oil were effective in reducing particulate emissions. For added versatility the organo-metallic compound is usually blended with dispersants and other ingredients to make a multipurpose product. Judicious application of these "combustion catalysts" should enable the user to cut fuel costs, while lowering air pollution at the same time. A partial table of contents follows here:

ISBN 0-8155-0729-1 **230 pages**